高等学校土木工程专业国际化人才培养英文系列教材

Engineering Geology

工程地质

Editor-in-Chief
Changbin Yan
闫长斌　主编

Associate Editors
Shuqian Duan　　Yunlong Liu
段淑倩　刘云龙　副主编

图书在版编目（CIP）数据

工程地质 = Engineering Geology：英文 / 闫长斌主编；段淑倩，刘云龙副主编 . —北京：中国建筑工业出版社，2023.7
高等学校土木工程专业国际化人才培养英文系列教材
ISBN 978-7-112-28779-6

Ⅰ.①工… Ⅱ.①闫… ②段… ③刘… Ⅲ.①工程地质—高等学校—教材—英文 Ⅳ.①P64

中国国家版本馆CIP数据核字（2023）第098896号

责任编辑：刘婷婷　吉万旺
书籍设计：锋尚设计
责任校对：姜小莲

高等学校土木工程专业国际化人才培养英文系列教材
Engineering Geology
工程地质
Editor-in-Chief
Changbin Yan
闫长斌　主编
Associate Editors
Shuqian Duan　Yunlong Liu
段淑倩　刘云龙　副主编

*

中国建筑工业出版社出版、发行（北京海淀三里河路9号）
各地新华书店、建筑书店经销
北京锋尚制版有限公司制版
建工社（河北）印刷有限公司印刷

*

开本：787毫米×1092毫米　1/16　印张：13½　字数：384千字
2023年6月第一版　　2023年6月第一次印刷
定价：46.00元（赠教师课件）
ISBN 978-7-112-28779-6
　　（41061）

版权所有　翻印必究
如有内容及印装质量问题，请联系本社读者服务中心退换
电话：（010）58337283　　QQ：924419132
（地址：北京海淀三里河路9号中国建筑工业出版社604室　邮政编码：100037）

Engineering Geology is a professional basic course for undergraduates majoring in civil engineering in colleges and universities, and an important supporting course aiming at value guidance, knowledge transfer and ability training for civil engineering majors. With the increasing international influence of Chinese engineering construction, the number of international undergraduate students majoring in civil engineering in China has increased significantly. This book intends to meet the needs of international education for Chinese civil engineering major, and at the same time improve the international integration ability of Chinese undergraduate students. The English text book *Engineering Geology* is written in accordance with the civil engineering teaching plan and requirements of Chinese colleges and universities, with reference to Chinese professional specifications and engineering experience.

Aiming at the training of civil engineering professionals under the background of the "Belt and Road" initiative, and focusing on understanding geological environment, serving engineering construction, preventing geological disasters, and protecting environment, this book emphasizes on cultivating students' ability to integrate theories with practice, based on a systematic explanation of the basic concepts and principles of engineering geology. The book has 8 chapters, and its main contents include: the definition, object and development history of engineering geology, the basic characteristics and physical-mechanical properties of minerals and rocks, geological structures and their influences on projects, engineering geological properties of rocks and rock masses, classification of soil and its engineering geological properties, groundwater and its engineering geological problems, adverse geological phenomena such as earthquakes, collapses, landslides, Karsts, and mudslide and their engineering geological problems, as well as engineering geological investigations. The content of this book covers all the main knowledge points listed in the "*Guiding Professional Specifications for Civil Engineering Undergraduates in Colleges and Universities*", and meets the requirements of the basic examination syllabus for registered civil (geotechnical) engineers.

This textbook can be used as a text book for the English courses or bilingual courses of ***Engineering Geology*** for civil engineering, urban underground space engineering, road bridge and river crossing engineering, railway engineering, civil water conservancy and ocean, civil water conservancy and transportation and other related majors. It can also be used as a reference for other related undergraduates, postgraduates and engineering technicians in geological engineering, geotechnical engineering, tunnel and underground engineering, hydraulic engineering, mining engineering, engineering management, etc.

《工程地质》是高等学校土木工程类专业本科生的一门专业基础课，是土木工程类专业开展价值引领、知识传授与能力培养的重要支撑课程。随着我国工程建设国际影响力的日益增强，土木工程类专业来华本科留学生规模大幅度增加。本书是为满足我国土木工程专业国际教育教学需求，同时提高我国本科学生国际接轨能力，根据中国高等学校土木工程类教学计划和要求，参考中国规范和工程经验编写而成。

本教材以"一带一路"背景下土木工程类专业人才培养为目标，以认识地质环境、服务工程建设、预防地质灾害和保护绿水青山为主线，在系统阐述工程地质基本概念与原理的基础上，注重培养学生理论联系实际的能力。全书共分为8章，主要内容包括：工程地质学科的定义、对象与发展历史，矿物与岩石的基本特征和物理力学性质，地质构造及其对工程的影响，岩石与岩体的工程地质性质，土的分类及其工程地质性质，地下水及其对工程建设的影响，地震、崩塌、滑坡、岩溶、泥石流等不良地质现象及其工程地质问题，以及工程地质勘察等内容。本书内容涵盖了《高等学校土木工程本科指导性专业规范》列出的全部主要知识点，满足注册土木工程师(岩土)基础考试大纲基本要求。

本教材可作为土木工程、城市地下空间工程、道路桥梁与渡河工程、铁道工程、土木水利与海洋、土木水利与交通等土木类专业及相关专业的《工程地质》英文课程或双语课程的教学用书，也可供地质工程、岩土工程、隧道与地下工程、水利工程、采矿工程、工程管理等相关专业本(专)科生、研究生和工程技术人员学习参考。

本书配备教学课件，可以向采用本书作为教材的老师提供。请有需要的任课教师按以下方式索取课件：1. 邮件：jckj@cabp.com.cn 或 jiangongkejian@163.com（邮件主题请注明《工程地质》英文版）；2. 电话：（010）58337285；3. 建工书院：http://edu.cabplink.com。

Preface
前言

Engineering Geology is a mandatory basic course for civil engineering majors, urban underground space engineering majors and geological engineering majors in colleges and universities. With the continuous expansion of the scale of engineering construction, there will be more complicated engineering geological conditions, and greater engineering geological problems and challenges in the practice of civil engineering and geological engineering. Therefore, it is particularly important to correctly understand the in-situ engineering geological conditions and comprehensively analyze existing engineering geological problems. Engineering geology organically combines general geology and basic knowledge of engineering construction, and plays a transition and cohesion role.

This book is characterized by the teaching of Chinese engineering geology courses and is compiled in accordance with the requirements of the *"Undergraduate Guiding Professional Code of Civil Engineering for Colleges and Universities"*. The main content has 8 chapters, including introduction, minerals and rocks, geological structure and its influence on engineering, classification and engineering geological properties of soil, engineering geological properties of rocks and rock masses, groundwater and its engineering geological problems, adverse geological phenomena and engineering geological investigations, etc.

The key characteristics of this book are as follows:

First, aiming at the training of civil engineering professionals under the background of the "Belt and Road" initiative, and focusing on understanding the geological environment, serving engineering construction, preventing geological disasters, and protecting environment, this book gives high priority to cultivating students' ability to integrate theories with practice, based on a systematic explanation of the basic concepts and principles of engineering geology. In addition, the book incorporates "ideological and political" elements, highlights the characteristics of "curriculum ideological and political construction", and focuses on cultivating students' ecological concepts, engineering awareness and craftsmanship.

Second, considering that civil engineering students do not have the basic professional courses such as geology foundation, this book introduces the basic theories and principles of general geology into the classroom, and combines them with engineering construction. It focuses on cultivating students' ability to analyze engineering problems and provides the basic knowledge for following courses, e.g., soil mechanics, rock mechanics, slope engineering, tunnel engineering and other professional courses.

Third, closely combining with the standards, specifications and engineering manuals currently in force in China, this book not only covers all the knowledge points of the *"Undergraduate Guiding Professional Code of Civil Engineering for Colleges and Universities"*, but also the requirements of the basic examination outline for registered civil (geotechnical) engineers in China. It could be used as exam preparation review materials as well.

Fourth, based on the teaching objectives of the course, the book condenses the teaching content into four topics: basic knowledge, geological structure and geological map reading, water and geological disaster and engineering geological investigations, which are conducive to the development of topic-based teaching.

Fifth, in the process of knowledge explanation, this book focuses on introducing engineering cases, combined with detailed pictures, video and other appropriate extended materials, so that learners can more directly understand the knowledge points.

It's recommended to use this book in the fourth semester for 32 class hours with field practice no less than 1 week.

This book is edited by Professor Changbin Yan, Associate Professor Shuqian Duan and Dr. Yunlong Liu from Zhengzhou University. Changbin Yan is editor-in-chief. Shuqian Duan and Yunlong Liu are associate editors. Changbin Yan wrote chapters 1, 3 and 6, Shuqian Duan wrote chapters 2, 4 , 7.4-7.6, and Yunlong Liu wrote chapters 5, 7.1-7.3 and 8, and Changbin Yan is the final editor.

In addition, graduate students e.g., Peiming Cheng, Yanyan Xia, Shipeng Jiang, Chongxuan Yuan, Po Gao, Minghuan Zhang, Yuanda Sun, Laibiao Xu have also participated in the text proofreading and editing work. They are appreciated for hard work.

Due to the limited level of editors, maybe there are some mistakes or shortcomings in the book. We welcome criticism and suggestions from readers. Some pictures and case materials in this book come from the Internet. We would like to express our high respect and gratitude to authors of the book references.

工程地质是高等学校土木工程专业、城市地下空间工程专业和地质工程专业必修的一门专业基础课程。随着工程建设规模的不断扩大，土木工程和地质工程实践活动中面临的工程地质条件越来越复杂，工程地质问题越来越多，挑战也越来越大。因此，正确认识工程活动场地的工程地质条件，全面分析存在的工程地质问题便显得尤为重要。工程地质将普通地质学和工程建设类专业基础知识有机结合起来，起到了承前启后的作用。

本书结合我国工程地质课程的教学特点，根据《高等学校土木工程本科指导性专业规范》要求编写，主要内容共分8章，包括绪论、矿物与岩石、地质构造、岩石与岩体的工程地质性质、土的分类与工程地质性质、地下水及其工程地质问题、不良工程地质现象及工程地质勘察等内容。本书的特点体现在：

（1）以"一带一路"背景下土木工程类专业人才培养为目标，以认识地质环境、服务工程建设、预防地质灾害和保护绿水青山为主线，在系统阐述工程地质基本概念与原理的基础上，注重培养学生理论联系实际的能力。此外，本书融入"思政"元素，突出"课程思政"特色，着重培养学生的生态理念、工程意识和工匠精神。

（2）考虑到土木专业学生未上过地质学基础等专业基础课程，本书将普通地质学的基本理论和基本原理引入课堂，并将其与工程相结合，注重培养学生分析工程问题的能力，同时为后续土力学、岩体力学、边坡工程、隧道工程等专业课程学习提供基础知识。

（3）紧密结合中国现行有关标准、规范及工程手册，除涵盖《高等学校土木工程本科指导性专业规范》的全部知识点外，还结合我国注册土木工程师（岩土）基础考试大纲的要求进行编写，可以作为备考复习资料使用。

（4）基于课程教学目标，本书将教学内容凝聚成四个专题：基础知识专题、地质构造与地质图阅读专题、水与地质灾害专题和工程地质勘察专题，有利于开展专题化教学。

（5）在知识讲解过程中，本书注重引入工程案例，配合图片资料和适当的扩展资料让学习者能更直观地理解知识点。

使用本书作为课堂理论教材时，建议修读学期为第四学期，建议学时为32学时，野外实习不少于1周学时。

本书由郑州大学闫长斌教授、段淑倩副教授和刘云龙博士编写。其中，闫长斌为主编，段淑倩和刘云龙为副主编。闫长斌编写第1、3、6章，段淑倩编写第2、4章和第7.4~7.6节，刘云龙编写第5章、第7.1~7.3节和第8章，最后由闫长斌统稿。此外，硕士研究生程沛铭、夏妍妍、蒋师朋、原崇轩、高坡、张明焕、孙远达、许来彪等也参与了部分文字校对和编辑工作，在此表示感谢。

限于编者水平有限，书中难免有错误与不足之处，恳请读者批评指正。本书部分图片和视频资料来源于网络。本书也参考了国内外许多前辈和同行的著作，在此对这些作者致以崇高的敬意并表示感谢。

Contents

Chapter 1   **Introduction**

    1.1 The Relationship between Engineering Activities and Geological Environment   002
    1.2 Definition, Objects and Tasks of Engineering Geology   005
    1.3 Engineering Geological Conditions and Problems   006
        1.3.1 Engineering Geological Conditions   006
        1.3.2 Engineering Geological Problems   007
    1.4 Research Methods of Engineering Geology   009
    1.5 Brief Description of the Development of Engineering Geology   009
    1.6 Learning Methods and Requirements of Engineering Geology   010
    Quiz   011

Chapter 2   **Minerals and Rocks**

    2.1 Earth and its Spherical Structure   014
        2.1.1 Inner Structure of Earth   014
        2.1.2 Geological Action   014
    2.2 Minerals   018
        2.2.1 Classification of Minerals   018
        2.2.2 Morphology of Minerals   019
        2.2.3 Physical and Mechanical Properties of Minerals   020
        2.2.4 Common Minerals and their Main Characteristics   023
        2.2.5 Preliminary Identification of Minerals   023
    2.3 Rocks   023
        2.3.1 Igneous Rock   023
        2.3.2 Sedimentary Rock   030
        2.3.3 Metamorphic Rock   036
        2.3.4 Comparison of Geological Characteristics of Different Rocks   040
    Quiz   042

Chapter 3   **Geological Structures**

    3.1 Geological Process   044
    3.2 Geological Time and Geochronological Scale   045
        3.2.1 Geological Time and its Determination Methods   045
        3.2.2 Geochronological Scale   048
    3.3 Attitude of Strata and Contact Relationships   048
        3.3.1 Attitude of Strata   048
        3.3.2 Determination and Expression Methods of Strata Attitude   048
        3.3.3 Outcrop Characteristics and Contact Relationships of Strata   049

3.4 Geological Structure: Folds — 053
  3.4.1 Conceptions — 053
  3.4.2 Types of Folds — 054
  3.4.3 Elements of Folds — 054
  3.4.4 Classification of Folds — 055
  3.4.5 Field Identification Methods of Folds — 057
  3.4.6 Engineering Significance of Folds — 058
3.5 Geological Structure: Fractures — 058
  3.5.1 Joints — 059
  3.5.2 Faults — 063
3.6 Geological Map — 069
  3.6.1 Concepts and Types of Geological Map — 069
  3.6.2 Representation Methods of Geological Conditions on the Geological Map — 069
  3.6.3 Geological Section Map and Comprehensive Geological Histogram — 070
  3.6.4 Reading and Analysis of Geological Map — 072
Quiz — 074

Chapter 4 Engineering Geological Characteristics of Rock and Rock Mass

4.1 Rock and Rock Mass — 076
4.2 Physical and Hydraulic Properties of Rock — 076
  4.2.1 Physical Properties of Rock — 076
  4.2.2 Hydraulic Properties of Rock — 076
4.3 Mechanical Properties of Rock — 078
  4.3.1 Deformation of Rock — 078
  4.3.2 Strength of Rock — 079
  4.3.3 Main Factors Affecting Rock Mechanical Properties — 080
4.4 Structure Types and Engineering Geological Evaluation of Rock Mass — 081
  4.4.1 Structural Plane — 082
  4.4.2 Structural Body — 088
  4.4.3 Structure Types of Rock Mass and Their Engineering Geological Characteristics — 089
Quiz — 091

Chapter 5 Classification and Engineering Geological Properties of Soils

5.1 Introduction — 094
5.2 Origin Types and Characteristics of Soil — 094
  5.2.1 Residual Soil (Q^{el}) — 094
  5.2.2 Deluvial Soil (Q^{dl}) — 094
  5.2.3 Diluvial Soil (Q^{pl}) — 095
  5.2.4 Alluvial Soil (Q^{al}) — 095
  5.2.5 Glacial Soil (Q^{gl}) — 095
  5.2.6 Aeolian Soil (Q^{aeo}) — 095
5.3 Composition and Structure of Soil — 096
  5.3.1 Solid Phase of Soil — 096
  5.3.2 Liquid Phase of Soil — 100
  5.3.3 Gas Phase of Soil — 101

	5.3.4 Structure of Soil	102
5.4	**Physical and Mechanical Properties of Soil**	104
	5.4.1 Three-phase Ratio Index of Soil	104
	5.4.2 Physical State Index of Non-cohesive Soil	107
	5.4.3 Physical State Index of Cohesive Soil	109
	5.4.4 Mechanical Properties of Soil	110
5.5	**Engineering Classification of Soil**	112
	5.5.1 By Accumulation Age	112
	5.5.2 By Geological Origin	112
	5.5.3 By Particle Size Distribution or Plasticity Index	112
	5.5.4 By Organic Matter Content	113
5.6	**Engineering Geological Properties of General Soil**	114
	5.6.1 Gravel Soil	114
	5.6.2 Sandy Soil	115
	5.6.3 Cohesive Soil	115
Quiz		115

Chapter 6 Groundwater and its Engineering Effects

6.1	**Basic Conceptions**	118
6.2	**Physical Properties and Chemical Composition**	118
	6.2.1 Physical Properties of Groundwater	118
	6.2.2 Chemical Composition of Groundwater	119
6.3	**Classification of Groundwater**	121
	6.3.1 Aquifer and Aquiclude	121
	6.3.2 Types and Characteristics of Groundwater	121
6.4	**Flowing of Groundwater**	128
	6.4.1 Basic Laws of Groundwater Movement	128
	6.4.2 Dewatering Problems in Foundation Pit Construction	131
6.5	**Impacts of Groundwater on Engineering Construction**	132
	6.5.1 Land Subsidence and Collapse	132
	6.5.2 Seepage Failure of Groundwater	132
	6.5.3 Floating Effect	136
	6.5.4 Inrushing of Foundation Pit	137
	6.5.5 Water Gushing and Mud Bursting in Underground Caverns	138
	6.5.6 Erosion of Groundwater	139
Quiz		142

Chapter 7 Adverse Geological Phenomena and Geological Problems

7.1	**Earthquake**	146
	7.1.1 Basic Conceptions	146
	7.1.2 Earthquake Destructive Effects and Seismic Fortification Measures	148
7.2	**Landslides**	150
	7.2.1 Basic Conceptions	150
	7.2.2 Classification of Landslides	151
	7.2.3 Formation Conditions of Landslides	152
	7.2.4 Developing Process of Landslides	154
	7.2.5 Field Identification of Landslides	155
	7.2.6 Prevention Measures and Treatments of Landslides	156

7.3 Dangerous Rock and Collapse 159
7.3.1 Basic Conceptions 159
7.3.2 Classification of Collapses 159
7.3.3 Formation Conditions of Collapse 161
7.3.4 Prevention Measures and Treatments of Dangerous Rock and Collapse .. 162
7.4 Mudslide 163
7.4.1 Basic Conceptions 163
7.4.2 Classification of Mudslide 164
7.4.3 Formation Conditions of Mudslide 166
7.4.4 Features of Mudslide 168
7.4.5 Prevention Measures and Treatments of Mudslide 168
7.5 Karst and Earth Cave 170
7.5.1 Karst 170
7.5.2 Earth Cave 174
7.5.3 Engineering Geological Problems of Karst and Earth Cave 175
7.5.4 Prevention Measures and Treatments of Karst and Earth Caves 175
7.6 Land Subsidence and Ground Collapse 176
7.6.1 Land Subsidence 176
7.6.2 Ground Collapse 178
7.6.3 Prevention Measures and Treatments of Ground Subsidence and Collapse 179
Quiz 180

Chapter 8 Engineering Geology Investigation
8.1 Grades and Stages of Engineering Geological Investigation 182
8.1.1 Investigation Grades 182
8.1.2 Investigation Stages 183
8.2 Methods of Engineering Geological Investigation 183
8.2.1 Engineering Geological Mapping 184
8.2.2 Exploration and Sampling 185
8.2.3 In-situ Testing Technology 188
8.2.4 Overview of Laboratory Tests 198
8.2.5 In-situ Inspection and Monitoring 199
8.3 Report of Engineering Geological Investigation 199
Quiz 200

References 201

Chapter 1
Introduction

1.1 The Relationship between Engineering Activities and Geological Environment

The geological environment is an extremely important part of the human environment. It mainly refers to the combination of the geological background, geological processes and their occurrence spaces which are closely related to the survival and development of human beings. It is also called the geological environment system.

On the one hand, the geological environment restricts engineering construction. For example: ① The earthquake zone restricts the urban layout. ② The landslide can force the project to be rerouted. ③ The soft soil foundation needs to be treated before construction, and so on. On the other hand, engineering construction will in turn affect the geological environment. For example: ① Extraction of groundwater can cause ground subsidence and ground collapse. ② The excavation of the road cutting can cause a landslide. ③ The construction of the bridge can cause erosion and silting deformation of some river sections.

(1) First, this is the influence of geological environment on engineering construction.

1) In many cases, the earthquake is ranked first.

For example:

① The Tangshan earthquake in 1976 was the worst earthquake disaster since the founding of the People's Republic of China. The death toll exceeded 240,000, second only to that of the Haiyuan earthquake in 1920, with 288,200 deaths (Fig. 1-1).

② The Wenchuan earthquake on May 12, 2008 reached a magnitude of 8.0 and the deaths toll were close to 70,000. Since then, "5·12" has been designated as the National Day for Disaster Prevention and Mitigation (Fig. 1-2).

2) In addition to earthquakes, landslides rank the second as a kind of geological environment disaster that has a greater impact on engineering construction.

Fig. 1-1 Comparison of new Tangshan and the ruined urban area after the earthquake

(a) before (b) after

Fig. 1-2 Beichuan County before and after the "5·12" earthquake

For example:

① On January 13, 2001, a magnitude 7.6 earthquake struck Salvador, Brazil, with the epicenter located 60 miles southwest of Santa Miguel. As a result, a landslide was caused in Santa Tecla, which eventually killed more than 700 people.

② On December 20, 2015, a large-scale landslide struck the Guangming New District of Shenzhen, China, resulting in 69 deaths and 8 missing.

3) Besides landslides, mudslides are also adverse geological phenomena that have a significant impact on engineering construction.

① On August 7, 2010, it rained heavily and suddenly in Zhouqu County, Gannan Tibetan Autonomous Prefecture, China, and a giant mudslide occurred in the county seat consequently (Fig. 1-3).

② On July 11, 2003, a mudslide occurred in Danba County, Ganzi Prefecture, Sichuan Province, China. One person was killed, 50

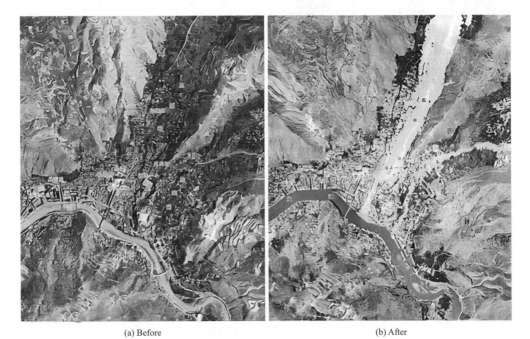

(a) Before (b) After

Fig. 1-3 Before and after Zhouqu debris flow

Fig. 1-4 "7·11" heavy mudstide in Qiongshan Gully, Danba County, Ganzi Prefecture, Sichuan Province, china

Fig. 1-5 Ground subsidence caused by excessive groundwater extraction

Fig. 1-6 Ground subsidence caused by metro shield construction

Fig. 1-7 Sunken Mexico City

people missing, and more than 1,230 people affected. Eight households with 88 houses collapsed and 55 houses were damaged. The mudslides washed away 1,000 m of provincial roads, 38 kilometers of mechanical farming roads, 6 bridges, 9.2 hectares of arable land, 20 hectares of non-cultivated land, and more than 1,000 meters of river embankment and 2 cars were buried (Fig. 1-4).

(2) Conversely, let's talk about the impact of engineering construction on geological environment with examples. The reform degree of the Earth's surface by human engineering activities has reached a level that cannot be ignored. So far, human activities have involved 80% of the surface area. The following pictures shows ground subsidence caused by excessive exploitation of groundwater (Fig. 1-5), and improper subway construction (Fig. 1-6), etc.

1) Ground subsidence resulted from dense population in cities. Mexico began to pump groundwater in 1850. From 1891 to 1973, the entire old city had sunk 8.7 m, causing damage to roads, buildings and other facilities on the ground. The drop of the groundwater level in the soil layer increases the effective stress, leading to further consolidation and settlement of the foundation (Fig. 1-7).

2) Petroleum extraction can also cause ground subsidence. The Wilmington Oilfield in Long Beach City, the United States had sunk 9 m from 1925 to 1977 (Fig. 1-8).

3) The construction of water conservancy projects will also affect the geological environment to a certain extent. Two persuasive examples are Malpasset arch dam (Fig. 1-9) and Hoover dam (Fig. 1-10).

Through the above cases, we have deeper understanding of the mutual influence of the geological environment and engineering construction. Human engineering activities and the geological environment are in a contradiction, in which they are interrelated and mutually restricted.

Fig. 1-8 Wilmington Oilfield District, Long Beach City, U.S.

Fig. 1-9 Malpasset arch dam on the Ravran River, France

Fig. 1-10 Impacts on the geological environment during the construction of the Hoover dam, USA

1.2 Definition, Objects and Tasks of Engineering Geology

Engineering geology is a marginal science formed by the mutual penetration and intersection of engineering science and geological science. It applies the principles of geology to the practice of civil engineering, through engineering surveys, investigations and research on the topography, stratum lithology, geological structure, engineering characteristics of rock and soil, hydrogeology and surface geology and other engineering geological conditions.

The ultimate goal of engineering geology is to seek a balance between engineering activities and the geological environment through the study the mutual constraints between human engineering activities and the geological environment, and rational development and proper protection of the geological environment.

1.3 Engineering Geological Conditions and Problems

1.3.1 Engineering Geological Conditions

Engineering geological conditions refer to the integration of the geological environment related to engineering activities, which include topography, stratum lithology (rock and soil types and properties), geological structures, hydrogeology, adverse geological effects (physical geological phenomena) and natural building materials, etc. It should be noted that we cannot take one of the above-mentioned aspects, but the whole of them as engineering geological conditions.

(1) Topography and landforms

Topography refers to the undulations of the ground surface, the steepness of the slopes, the width and morphological characteristics of the valleys, etc. The landforms refer to the reason, process and time of the formation of the topography. The undulations, thick soil layers and outcropping of bedrock, groundwater burial conditions and surface geological phenomena in plain, hilly, and mountainous regions have different characteristics. These factors can all directly affect the choice of construction sites and routes.

(2) Stratum lithology

The rock and soil that make up the stratum are the foundation of buildings, building material or medium. Its characteristics include genesis, age, lithology, attitude, diagenesis characteristics, degree of metamorphism, weathering characteristics, weak interlayers and contact zones, and physical and mechanical properties. The stratum lithology has a stake in the safety and stability of the building, so it is the most basic engineering geological factor.

(3) Geological structures

Geological structures include the distribution and characteristics of folds, faults, and joint structures, especially the formation of new era and large-scale active faults. Such structures have a control effect on earthquakes and other disasters, and are therefore of great significance to the safety and stability, settlement and deformation of buildings. Therefore, geological structure is the basic object of engineering geological research.

(4) Hydrogeological conditions

Hydrogeological conditions include the origin, store, distribution, dynamics and chemical composition of groundwater. Groundwater is an important factor that can reduce the stability of rock and soil mass, and special attention should be given to it in engineering construction. For example, the change of groundwater table should be considered in the calculation of foundation bearing capacity and settlement. In order to prevent foundation pit gushing and flowing soil in foundation pit, piping of embankment and road boiling, etc., the first thing should be taken into account is the change of groundwater table. In addition, the corrosiveness of groundwater to building materials is another engineering problem that we have to face.

(5) Surface geology

Surface geology reflects modern surface geological action. It is closely related to the topography, climate, lithology, structure, groundwater and surface water of the construction area. Surface geology mainly includes landslides, collapses, karsts, mudslides, sand movement, river erosion and deposition. It can affect the overall layout, design and construction methods of buildings, and is of great significance for evaluating the stability of buildings and predicting changes in engineering geological conditions.

(6) Natural building materials

Natural building materials refer to soil and stone materials for construction. In order to save transportation costs, the principle of "obtaining raw material locally" should be followed. Therefore, natural building materials have a greater impact on the cost of the project.

1.3.2 Engineering Geological Problems

Engineering geological problems refer to geological problems that have a major impact on the design, construction, and operation of projects caused by the interaction and contradiction between engineering construction and the geological environment. Specifically, it refers to the problems that often occur in the stability, economy or normal use of buildings due to the fact that the engineering geological conditions of the study area cannot meet the needs of a certain engineering construction. It mainly includes two major issues: one is the problem of regional stability, and the other is the problem of engineering stability, such as foundation stability, slope stability, and stability of tunnel surrounding rocks.

(1) Regional stability

Engineering geological problems arising from specific geological conditions and affecting large areas include active faults, earthquakes, reservoir-induced earthquakes, seismic sand liquefaction and land subsidence, etc. Mastering the regularity of these issues is of great significance to the planning and site selection or the rational development and protection of the geological environment.

(2) Engineering stability

1) Foundation stability

The settlement and deformation of different sizes will be generated due to the load of the superstructure of the foundation. Excessive or uneven settlement and deformation will cause cracks, tilt, and collapse of the buildings, thus affecting normal use, and even causing instability and collapse.

Lack of attention to stability of the foundation may lead to serious consequences, such as the case of the famous Canadian Transcon barn foundation instability (Fig. 1-11).

The barn is 59.4 m in length, 23.5 m in width, and 31.0 m in height, with 65 silos. The barn was underpinned by reinforced concrete raft foundation with its thickness of 61 cm, and the buried depth 3.66 m, and a weight of 20,000 t. Its construction started in 1911 and completed in 1913, when a serious accident occurred. The grain was loaded in September, 1913, and 31,822 t grain was loaded on October 17, on which day the vertical settlement reached 30.5 cm in one hour. The 24-hour tilt 26°53′, the west end sunk 7.32 m, the east end rose 1.52 m and the upper steel concrete silo were intact however.

The cause of the accident was that the foundation soil was not investigated prior to construction. According to the results of the

Fig. 1-11 The instability of Canada's Transcon barn due to foundation problems

excavation and borrowing test of the adjacent structure foundation trench, the foundation bearing capacity was calculated for this barn. After investigation and calculation in 1952, the actual bearing capacity of the foundation was less than the foundation pressure when accident occurred. Therefore, the barn foundation slipped due to tensile rupture led by overload.

The corrective measures were using more than 70 concrete piers to support the bedrock, 38,850 t jacks and supporting systems to gradually correct the barn body, and it descended by 4 meters (Fig. 1-12).

2) Slope stability

Excavating or filling artificial slopes (cuttings, embankments, dams, foundation pits, etc.) is necessary in engineering activities, especially in road engineering. Slope stability is very important to prevent geological disasters and ensure foundation stability.

The famous Vajont dam failure accident happened in the night of October 1963 is a convincing example to illustrate slope stability. A large landslide occurred on the left bank of the Vajont reservoir in the mountains of northern Italy (Fig. 1-13). The flood destroyed five villages within a several kilometers downstream, and 1,925 people died in their dreams. Vajont dam failure is recognized as the most serious landslide disaster in the world. The cause of the landslide accident was water storage.

3) The stability of the underground cavity

The underground cavities are surrounded by the rock-soil medium (crown arch, wall rock, etc.), and the original equilibrium condition of the underground rock mass can be destroyed during the excavation and construction of the chamber. This will cause a series of unstable phenomena like wall rock collapse, and groundwater piping.

4) Groundwater problems

The corrosion of concrete by groundwater is a complex physical and chemical process.

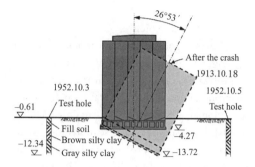

Fig. 1-12 Investigation and correction of the barn foundation

Fig. 1-13 A large-scale landslide occurred on the left bank of the Vajont Reservoir in the mountains of northern Italy

Under certain engineering geological and hydrogeological conditions, it has a great impact on the durability of building materials.

In addition, in the morning of June 10, 2018, a water inrush incident occurred at the K170+671 Chaoyang tunnel of Guinan Railway, China. As of 14:30 in the afternoon of the day, two construction workers died on site, and one worker died after the rescue failed.

1.4 Research Methods of Engineering Geology

The research object of engineering geology is a complex geological body, so its research method should be a comprehensive one combining qualitative and quantitative analysis, specifically natural history analysis, mathematical mechanics, engineering geological analogy, and model simulation test method.

There are four main research methods of engineering geology:

(1) Natural history analysis

It is a qualitative study that analyzes the formation and development of engineering geological conditions from the perspective of geology and natural history.

Such method is applied in analysis on deformation and failure of slopes is a good example.

(2) Mathematical mechanics analysis

It is a quantitative study based on the research of the formation mechanism of engineering geological problems. It uses mechanical means to establish a model for calculation and prediction.

Examples include rigid body limit equilibrium method, theory of elasticity and plasticity, numerical analysis method, etc.

(3) Model simulation test

It obtains the physical and mechanical parameters of rock and soil through lab or field tests. Simulation experiments include photo-elastic experiment, electrical network simulation, network simulation of structure, etc.

(4) Engineering geology analogy

The engineering geological problems are analyzed based on the existing data in similar conditions, such as expert judgment, and empirical parameters.

For the engineering geology research methods, the University of British Columbia once proposed the famous UBC triangle model (Fig. 1-14). The method given by this model is a combination of geological model-indoor and outdoor tests-analysis. In fact, it is another understanding of the above four methods.

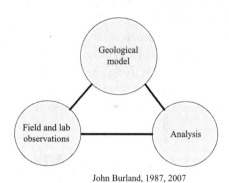

John Burland, 1987, 2007

Fig. 1-14 UBC triangle model

1.5 Brief Description of the Development of Engineering Geology

Engineering geology is an applied subject both ancient and young. The people in ancient times built many large-scale projects, and left us amazing projects. Some famous examples are as follows, the world-famous Great Wall, Dujiangyan irrigation project of China (Fig. 1-15), and the Pyramid of Egypt, etc.

Some buildings have been operating for more than one thousand years without breaking down and still exist today. The ancient people not only possessed superb architectural skills, but also had a considerable understanding of the engineering geological environment of the construction site.

Fig. 1-15 The Great Wall and Dujiangyan irrigation project

However, their knowledge about engineering geological environment is only confined to practice experience, and failed to form a systematic engineering geology and engineering construction theory due to limited productivity level at that time. Engineering geology has only more than 80 years of history as an independent branch of geology in the world.

In China, the development of engineering geology basically began in the 1950 s. In 1952, China established the Ministry of Geology and Mineral Resources, and the Bureau of Hydrogeology and Engineering Geology. At the same time, water conservancy, railway, transportation and other departments have successively set up engineering geological offices or survey teams.

From 11th October to 15th October in 2019, the National Engineering Geology Annual Conference was held in China University of Geosciences in Beijing, with more than 1300 representatives attended. From 16th October to 18th October in 2020, the 11th National Engineering Geology Congress was held in Wuhan both online and offline, with 340 on-site participants and 200,000 online live viewers. From 14th October to 17th October in 2021, the National Engineering Geology Annual Conference was held in Qingdao with over 100,000 attendees online. It can be seen that the development of engineering geology is vigorous and prosperous.

1.6 Learning Methods and Requirements of Engineering Geology

Engineering geology is a basic technical course for the students in the majors of civil engineering. Generally, it is offered before the courses of soil mechanics, rock mechanics and other professional courses. The course is characterized by wide contents, numerous concepts and practices. In order to learn well, it's better to fully understand those concepts, analysis reasonably, integrating theories with practice than simply reciting.

The requirements for this course for civil engineering and urban underground space engineering students are as follows: ① be familiar with the basic concepts of engineering geology. ② master basic engineering geological analysis methods. ③ know well the formation conditions and prevention measures of different engineering geological problems. ④ realize the inherent relationship among key geological characteristics, e.g., stratum lithology, geological structure, groundwater, unfavorable geological conditions, and engineering investigation methods.

Quiz

1. How to understand the relationships among engineering geology, geological engineering and geotechnical engineering?

2. What is the engineering geological conditions? What are engineering geological problems?

3. How do engineering geological conditions restrict engineering construction? How does engineering construction affect engineering geological conditions?

4. Please take a project as an example to illustrate the basic requirements of the engineering construction on the engineering geological environment.

5. What are the analysis methods of engineering geology?

Chapter 2
Minerals and Rocks

2.1 Earth and its Spherical Structure

2.1.1 Inner Structure of Earth

The Earth consists of two parts, i.e., the outer layer and the solid Earth. The outer layer is composed of the atmosphere, surrounding water, and the biosphere. Similarly, the solid Earth is composed of the Earth crust, Earth mantle, and Earth core, as shown in Fig. 2-1. The Earth mantle is 2900 km deep under the Earth's crust. The part under the Earth mantle is called the Earth's core, whose density is 11~16 g/cm^3, made up of material including Fe, Ni. The Earth's core is made up of the solid nuclear inside and the liquid state outside.

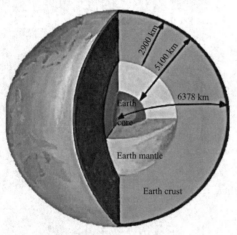

Fig. 2-1 Circular layers of the Earth

The knowledge about the Earth's internal strata mainly comes from the study of seismic waves. According to researches, there is a low-speed seismic wave at the top of the Earth mantle, in which 5% of the material is in a molten state called asthenosphere. The solid layer on the asthenosphere is the lithosphere. The lithosphere has stronger rigidity and could be split into many fragments called plates. The movement of plates on the asthenosphere is called plate movement, which is the origin of the tectonic movement.

Plate indicates the discontinuous plate-form rock blocks of lithosphere divided by mid ocean ridges, oceanic grooves, and transportation faults. According to Xavier Le Pichon, there are six plates, i.e., Eurasian plate, African plate, Pacific plate, American plate, Antarctic plate, and Indian Ocean plate.

Plate movement generally refers to the relative movement of one plate on the Earth's surface to another. With the movement of the asthenosphere, each plate will also move horizontally. The plate moves above the asthenosphere and is driven by the mantle convection. There are three relative motion modes between plates, i.e., aggregation, tension and dislocation. Hence, the boundary of plates could be divided into three types accordingly, i.e., aggregation plate boundary, tension plate boundary, and dislocation plate boundary (Fig. 2-2).

The basic materials of the Earth crust are various chemical elements, including O, Si, Al, Fe, Ca, Na, K, Mg and H which account for 98.13% of the total mass of the crust. Minerals are natural elementary substance or chemical compounds formed by geological action, and they have not only fixed chemical compositions, but also fixed inherent physical properties. Minerals are stable under some physical and chemical conditions. Rocks are compounds of one or more kinds of minerals, and the key components of Earth lithosphere.

2.1.2 Geological Action

Geological action means the action that makes alteration for the composition, structure, shape of the rock cycle because of the natural dynamics. The classification of the geological action is shown as follows.

(1) Outer-dynamics geological action

Outer-dynamics geological action refers to dynamics acting on earth surface from atmosphere, water and creatures accused by solar

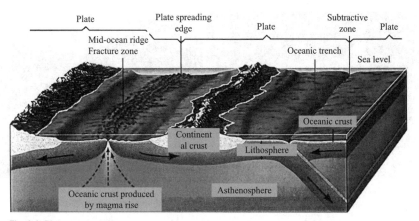

Fig. 2-2 Plate movement

power and gravity. Outer-dynamics geological action includes weathering, erosion, carrying, sedimentation and the rock-formation action.

1) Weathering (Fig. 2-3a). The rock is destroyed at the place where it was formed because of the outer dynamics. It mainly includes physical weathering caused by sunshine, temperature, wind and water, chemical weathering caused by chemical reaction and biological weathering caused by plant growth, insects, and microorganism.

2) Erosion (Fig. 2-3b). Rock body is separated from its mother rock because of the water, wind and glacier.

3) Carrying (Fig. 2-3c). In this process, rock and erosion materials are carried from one place to another by geological forces including traction transport, suspension transport and solution transport.

4) Sedimentation (Fig. 2-3d). In the process of carrying, the carried materials deposit when the weight of the materials is bigger than carrying force. Sedimentation contains mechanical sedimentation, chemical sedimentation and biological sedimentation.

5) Rock-formation action. In this process, the materials deposit to become hard or solid rock body due to the consolidation effect, cementation effect and re-crystallization effect.

(2) Inner-dynamics geological action

As for the inner-dynamics geological action, the dynamic is from the Earth. Such geological processes mainly occur in the Earth's interior, including crust movement, magmatism, earthquake and metamorphism.

Crust movement is the most important inner dynamics geological action among the four inner-dynamics geological actions, including lifting movement and horizontal movement. Lifting movement is relatively slow, during which some area in earth crust would be lifted and the other would subside. This can be also seen as orogenic movement, for example, the formation of Himalaya Mountains (Fig. 2-4a). Horizontal movement, including plate collision and cracking, is more violent than lifting movement. The Great Rift Valley, known as the biggest scar on earth, is

(a) Weathering (b) Erosion (c) Carrying (d) Sedimentation

Fig. 2-3 Outer-dynamics geological actions

(a) Himalaya Mountains

(b) Great Rift Valley

Fig. 2-4 Crust movement

just formed in the tension crack boundaries (Fig. 2-4b). The lifting movement and the horizontal movement are interactive and inseparable.

The magmatism mainly contains two types, i.e., the volcanic eruption and the magmatic intrusion (Fig. 2-5). Magma erupts or overflows to the ground, then condenses to rock. Magmatic intrusion means that the magma that does not erupt or overflows to the ground condenses to hypabyssal (shallow intrusive) rock and abyssal (deep intrusive) rock at different depth.

Earthquake means rapid crustal vibration. It can be seen from Fig. 2-6 that the place where earthquakes occur in the earth's crust is called the hypocenter. And point on the Earth's surface that is directly above the hypocenter is the epicenter, the distance from the hypocenter to the epicenter is called hypocenter depth. We take the distance between any point on the earth to epicenter as epicenter distance.

Metamorphism refers to the process that the rock is changed into a new kind of rock called metamorphic rock due to high pressure and temperature. Metamorphism is subjected to temperature, pressure and chemically active fluid. According to the influence factors, the metamorphism can be divided into three types of actions, including contact metamorphism, dynamic metamorphism and regional metamorphism. The contact metamorphism can be divided into contact thermal metamorphism and contact metasomatic metamorphism. Contact thermal metamorphism result from thermal flow with high temperature, and contact metasomatic metamorphism, usually partial but not general, occurs as interaction is

(a) Volcanic eruption

(b) Magmatic intrusion

Fig. 2-5 Magmatism

generally limited near the metamorphic zone with high temperature and low pressure, as shown in Fig. 2-7 (a). Dynamic metamorphism, related with faults, mostly happens in the crust active zones. As can be seen in Fig. 2-7(b), the particles become directional after the stress action. Particles in rock get stretched by shear. Regional metamorphism is caused by regional structure movement in large area (Fig. 2-7c).

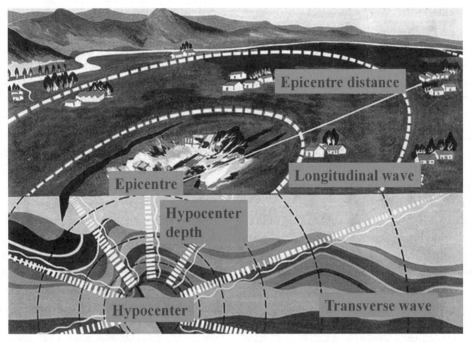

Fig. 2-6 Elements of earthquake

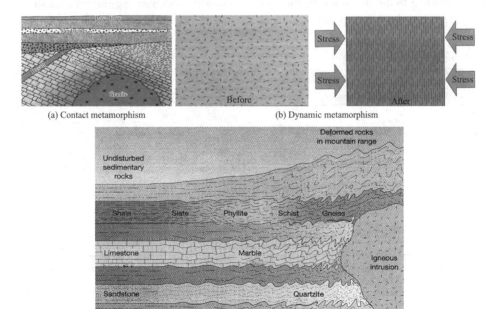

Fig. 2-7 Metamorphism

2.2 Minerals

Minerals are the basic substances that form rocks, and natural aggregates composed of one or more minerals in a certain pattern are called rocks.

Minerals are natural elements or compounds formed by a combination of one or several elements by geological actions, with relative fixed chemical composition and inner structure.

2.2.1 Classification of Minerals

The major classification of minerals is based on their chemical composition and chemical properties of compounds. It can be divided into simple substance, oxide, hydroxide, halide, sulfate, carbonate, phosphate and silicate. Each type of minerals has similar chemical and physical properties.

From the perspective of crystallography, it can be divided into crystalline minerals and non-crystalline minerals. Crystalline minerals can be divided into islands, rings, chains and layers according to the spatial arrangement of crystal particles (Fig. 2-8). Crystalline mineral is made up of crystalline substance, in which molecules are arranged regularly periodically in space, including phanerocrystalline and cryptocrystalline minerals. The non-crystalline minerals, whose molecules are not arranged regularly and periodically in space, consist of vitreous mineral and colloid mineral.

Minerals in nature can be divided into three categories according to their genetic types, i.e., primary minerals, secondary minerals and metamorphic minerals. Primary minerals, including quartz, feldspar, pyroxene, hornblende, mica, olivine, garnet, are generally formed by condensation of magma. Some secondary minerals, including kaolinite, montmorillonite, illite and chlorite, are generally generated directly by the weathering of primary minerals. And some secondary minerals, including calcite, gypsum, dolomite, are precipitated in aqueous solution.

However, there are only about 20~30 minerals that can influence the characteristics of rocks and are useful for rock identification. They are called rock-forming minerals. The common rock-forming minerals are shown in Table 2-1.

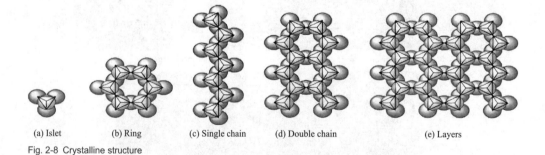

(a) Islet (b) Ring (c) Single chain (d) Double chain (e) Layers

Fig. 2-8 Crystalline structure

Common rock-forming minerals Table 2-1

Fe-rich magnesium silicate	Fe-poor magnesium silicate	Other minerals
Olivine, pyroxene, amphibole, garnet, serpentine, chlorite, biotite	Quartz, potassium feldspar, plagioclase, Muscovite, sericite, kaolinite, andalusite	Rutile, magnetite, limonite, pyrite, galena, dolomite, apatite, nepheline, calcite, etc.

2.2.2 Morphology of Minerals

The morphology of minerals is the shape of mineral monomers and aggregates. The morphology of minerals is controlled by its internal structure, chemical composition and formation environment. Monomer form refers to the crystal shape of a single mineral, while aggregate form refers to the combined form formed by the agglomeration of the same kind of minerals. Minerals are characterized by a common unique crystal morphology which is an important factor for the identification of minerals.

(1) Monomer morphology of crystals (Fig. 2-9)

1) One-way extension. Crystals grow and develop rapidly along an axis, while grow comparatively slow along the other direction. As a result, the crystals could form long strip, long columnar, and even acicular and fibrous shapes, and become minerals like quartz, amphibole, and asbestos.

2) Two-way extension. The crystal is plate-or sheet-shaped, such as gypsum, mica, chlorite, etc.

3) Three-way extension. The crystal is almost equiaxed and granular, i.e., cube, tetrahedron, and rhombohedron. Representative minerals of this type are octahedral magnetite, rhomboid dodecahedral garnet, etc.

(2) Aggregate morphology of mineral

Crystalline minerals rarely appear as monomers in nature, while amorphous minerals have no regular monomer morphology at all. Minerals are often identified by the morphology of aggregates. Common forms of mineral aggregates are shown in Fig. 2-10.

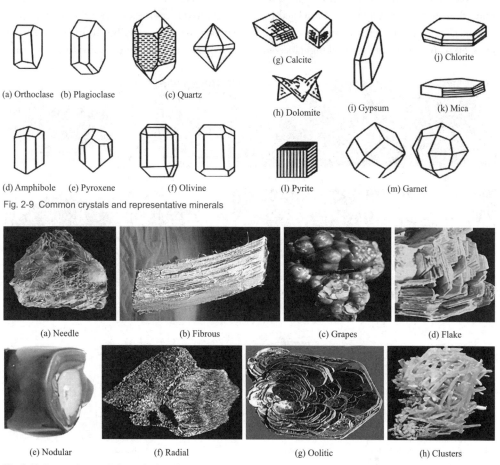

(a) Orthoclase (b) Plagioclase (c) Quartz (g) Calcite (j) Chlorite
(h) Dolomite (i) Gypsum (k) Mica
(d) Amphibole (e) Pyroxene (f) Olivine (l) Pyrite (m) Garnet

Fig. 2-9 Common crystals and representative minerals

(a) Needle (b) Fibrous (c) Grapes (d) Flake
(e) Nodular (f) Radial (g) Oolitic (h) Clusters

Fig. 2-10 Aggregate morphology of mineral

2.2.3 Physical and Mechanical Properties of Minerals

The physical properties of minerals include optical properties (i.e., color, streak, gloss and transparency), mechanical properties (i.e., hardness, cleavage and fracture), magnetism and electrical conductivity, etc. They are the main distinguishing features of minerals.

(1) Color

The color of a mineral refers to the color of the fresh surface when it exposed to natural light. The color depends on the chemical composition of the mineral and the impurities it contains. According to the cause of color, there are idiochromatic color, allochromatic color and false color (Fig. 2-11).

1) Idiochromatic color

Idiochromatic color is the inherent color of minerals, which is relatively fixed, and related to the chemical composition and crystal structure of the mineral. In the case of rock-forming minerals, the color varies greatly due to the complexity of the composition.

2) Allochromatic color

Allochromatic color refers to the color caused by some impurities in minerals, and the nature of the mineral itself is independent. Allochromatic color is not fixed and varies with the different impurities.

3) False color

False color is caused by the refraction and scattering of light by the cracks in the mineral interior or the oxide film on its surface.

(2) Streak

The color of the powder when the mineral is scratched on the white unglazed porcelain plate is called streak (Fig. 2-12). Streak, often used in mineral identification, can eliminate false color, and weaken allochromatic color.

(3) Luster

Luster is the ability of mineral fresh surface to reflect visible light, which is an important sign to identify minerals. According

Cinnabar (red)　　Sphalerite (orange)　　Celestite (syan)　　Azurite (blue)

Natural gold (golden)　　Talc (gteen)　　Corundum (purple)　　Cassiterite (black)

Fig. 2-11 Colors of minerals

Orpiment　　Golden　　Haematite　　Cherry red　　Chalcopyrite　　Black

Crocoisite　　Yellow　　Cinnabar　　Red　　Molybdenite　　Grey

Fig. 2-12 Mineral streaks

to luster strength degree, it can be divided as metallic luster, semi-metallic luster and nonmetallic luster. The reflection of metallic luster is as strong as the bright and dazzling plating of metal surface. A semi-metallic luster, weaker than a metallic luster, resembles an unpolished surface of iron. Nonmetallic luster indicates that the surface of the mineral has a weak ability to reflect light and it is an inherent feature of most nonmetallic minerals such as quartz and talc (Fig. 2-13).

Due to differences in mineral surface or the ways of mineral aggregates, the nonmetallic luster could show the following different characteristics.

1) Adamantine luster. The reflection of mineral surface is as strong as diamond.

2) Glass luster. The reflection of the mineral surface is similar to that of glass, such as the luster on the cleavage surface of feldspar and calcite.

3) Grease luster. The surface of the mineral appears to be coated with grease, like the luster on a quartz fracture.

4) Pearl luster. The mineral surface is like the luster of the nacre layer inside the shell, like the luster of mica.

5) Silk luster. The mineral surface is like a silky reflection, like the luster of gypsum.

6) Waxy luster. It is a luster on the surface of a dense mineral, such as the luster of serpentine or talc.

7) Earthy luster. The mineral surface is rough, dull and dim as soil, such as the luster of kaolinite.

(4) Transparency

Transparency refers to the degree of light going through the mineral, which is related to the ability of the mineral to absorb visible light. It can be divided into three degrees, i.e., transparent ability (i.e., crystal), translucent ability (i.e., cinnabar) and opaque ability (i.e., pyrite), as shown in Fig. 2-14.

(5) Hardness

Hardness refers to the ability of a mineral to resist scratching, pressing and grinding from external forces, which is an important sign indicating the hardness or softness of a rock. Different minerals have different hardness due to different chemical composition and internal structure, hence, hardness is also an important characteristic to identify minerals. The identification of mineral hardness should be carried out on the fresh crystal face or cleavage

Fig. 2-13 Mineral luster

(a) Crystal　　　　　　　　　　(b) Cinnabar　　　　　　　　　　(c) Pyrite

Fig. 2-14 Transparency

face of the mineral. The hardness could be divided into 10 levels, called Mohs hardness scales. The characteristic minerals of Mohs hardness are shown in Table 2-2.

In field investigation, nails (2~2.5), pencil sharpener (5~5.5), glass (5.5~6), steel blade (6~7) are commonly used to identify the hardness of minerals.

(6) Cleavage and fracture

Cleavage is the mechanical property that the mineral can be regularly broken into a plane along a certain direction under external mechanical force (such as beating and extrusion), and the plane is called cleavage plane.

Different crystalline minerals are different in the degree of cracking, the number of cleavage and the degree of complete cleavage plane because of their different internal structures. According to the number of cleavage directions, cleavage can be divided into one-way cleavage (i.e., mica), two-way cleavage (i.e., feldspar), three-way cleavage (i.e., calcite) and multi-way cleavage (i.e., fluorite) (Fig. 2-15).

According to the degree of completeness of cleavage, cleavage can be divided into the following four types (Fig. 2-16).

(1) Extreme complete cleavage. Such mineral, like mica, is easy to split into flakes with its cleavage surface large and complete, smooth and bright.

(2) Complete cleavage. The cleavage plane is often broken into small pieces, and the cleavage plane is not large, and the fracture is not easy to occur, which is often found in calcite.

(3) Medium complete cleavage. The cleavage surface is small and not smooth, and the fracture is more likely to appear, which is often found in feldspar and hornblende.

(4) Incomplete cleavage. Cleavage plane can be rarely seen under the action of external forces, such as the case in apatite and garnet.

Common fractures mainly include conchoidal fracture like a shell, even fracture, ragged fracture and uneven fracture, as shown in Fig. 2-17.

The cleavage and fracture of minerals can interact with each other. When cleavage is complete, no fracture appears. On the contrary, when the cleavage is incomplete or absent, the fracture can be significant. Such as quartz without cleavage, only conchoidal fracture appears (Fig. 2-17a).

Hardness tester by Mohs Table 2-2

Hardness	1	2	3	4	5	6	7	8	9	10
Mineral	Talc	Gypsum	Calcite	Fluorite	Apatite	Orthoclase	Quartz	Topaz	Corundum	Diamond

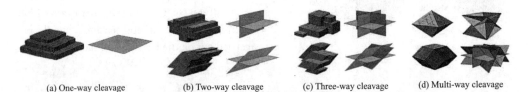

(a) One-way cleavage　　(b) Two-way cleavage　　(c) Three-way cleavage　　(d) Multi-way cleavage

Fig. 2-15 Cleavage classification according to the number of cleavage directions

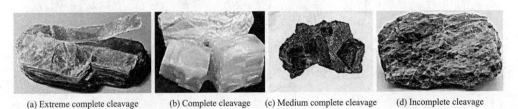

(a) Extreme complete cleavage　　(b) Complete cleavage　　(c) Medium complete cleavage　　(d) Incomplete cleavage

Fig. 2-16 Cleavage classification according to the degree of completeness

(a) Conchoidal fracture (b) Even fracture (c) Ragged fracture (d) Uneven fracture

Fig. 2-17 Fractures classification

2.2.4 Common Minerals and their Main Characteristics

The common minerals and their main characteristics are shown in Table 2-3, among which kaolinite, montmorillonite and illite are three common clay minerals to constitute clay rock and soil. These minerals, featured by small particles and colloidal properties, can have active physical and chemical interaction with water, resulting in complex and varied engineering geological characteristics.

2.2.5 Preliminary Identification of Minerals

Accurate identification of minerals requires a variety of methods in the laboratory, such as blowpipe analysis, differential thermal analysis, spectral analysis, chemical analysis, polarizing microscope analysis, electron microscope scanning, X-ray analysis. However, the most basic one is to identify minerals with the naked eyes or only the help of simple tools such as knives, porcelain plates, and magnifying glasses, so as to observe and preliminarily identify the appearance of minerals. Attention should be paid to its color, crystal shape, luster, hardness, cleavage and other main characteristics. The formation conditions of minerals and their symbiotic minerals should also be considered. The naked-eye-identified characteristics of common rock-forming minerals are shown in Table 2-3.

2.3 Rocks

Rock is composed of one or more minerals or other rock and mineral debris in geological process with a certain texture and structure characters of mineral assembly, and it is the basic unit of the crust.

The engineering geological characteristics of rocks are closely related to their mineral composition, texture and structure characters. Mineral composition, texture and structure characters are also the main basis for the identification of rocks. Rock texture refers to the degree of crystallization, particle size, shape, and combination of minerals in a rock. Rock structure refers to the arrangement and filling method among mineral aggregates or between mineral aggregates and other components in the rock. According to the geological origin, rocks can be divided into three categories, including igneous rock, sedimentary rocks, and metamorphic rocks.

2.3.1 Igneous Rock

Igneous rocks are cooled or solidified magma when it rises up into the crust along tectonically weak zones or erupts onto the ground from deep in the Earth's crust. Magma is a kind of silicate melt under high temperature (above 1,000) and high pressure deep in the

Main characteristics of common minerals

Table 2-3

No.	Mineral name	Composition	Hardness	Form	Color	Streak	Luster	Relative density	Cleavage or fracture
1	Talc	$Mg_3[Si_4O_{10}][OH]_2$	1	Plate, Sheet, Block	White, light green, light red	White	Glass, Waxy	2.7~2.8	Cleavage in one direction
2	Kaolinite	$Al_4(Si_4O_{10})[OH]_8$	1~2	Earthy and Lumpy	White			2.58-2.61	Soil fracture
3	Montmorillonite	$(Al_2Mg_3)(Si_4O_{10})[OH]_2$	1~2	Earthy, Micro-scaly	White, Off-white			2~3	Soil fracture
4	Illite (Hydromica)	$KAl_2[(Al, Si)Si_3O_{10}](OH)_2 \cdot nH_2O$	1~2	Earthy and Scaly	White		Lumpy, Grease luster		Soil fracture
5	Gypsum	$CaSO_4 \cdot 2H_2O$	2	Plate, Block, Fiber	White, Light gray	White	Glass, Pearl	2.6~2.9	Cleavage in one direction,
6	Chlorite	$(Mg, Al, Fe)_6[(Si, Al)_4O_{10}][OH]_8$	2~3	Scaly	Green		Pearl	2.3	Cleavage in parallel lamellar directions
7	Muscovite	$KAl_2[AlSi_3O_{10}][OH]_2$	2.5~3	Plate, Scaly, Aggregate	Colorless	White	Glass, Pearls	2.6~3.3	Cleavage in one direction
8	Biotite	$K(Mg, Fe)_3[AlSi_3O_{10}][OH]_2$	2.5~3	Short columnar, Plate, Flaky aggregate	Black, Brown,	Light green	Glass, Pearls	2.6~3.12	Cleavage in one direction
9	Serpentine	$Mg_6[Si_4O_{10}][OH]$	2.5~3.5	Fine scaly, Compact massive			Grease, Silk	3.02~3.12	
10	Calcite	$CaCO_3$	3	Rhomboid, Granular, Tuberculosis, Stalactite	Colorless, Gray	White	Glass	2.83	Three groups of cleavage
11	Dolomite	$CaMg(CO_3)_2$	3.5~4	Rhomboid, Granular, Massive	White, Light yellow, Red	White	Glass	2.6~2.8	Three groups of cleavage

Chapter 2 Minerals and Rocks

Continued

No.	Mineral name	Composition	Hardness	Form	Color	Streak	Luster	Relative density	Cleavage or fracture
12	Limonite	$Fe_2O_3 \cdot nH_2O$	5~5.5	Massive, Earthy, Beanlike, Honeycombed	Brown, Black	Fawn	Half metallic	3~4	No cleavage
13	Hematite	Fe_2O_3	5.5~6	Massive, Ridged, Oolitic	Steel gray, Iron black, reddish brown	Cherry red	Half metallic	5~5.3	No cleavage
14	Hornblende	$Ca_2Na(Mg,Fe)_4(FeAl)[(Si,A)_1O_{11}][OH]_2$	5~6	Long cylindrical, the cross section is hexagonal	Dark green to Black	Light green	Glass	3.1~3.3	Two groups of cleavage
15	Pyroxene	$Ca(Mg,Fe,Al)[(SiAl)_2O_6]$	5.5~6	Short columnar, the cross section is the octagon	Dark green	Celadon	Glass	3.23~3.56	Two groups of cleavage
16	Orthoclase	$K[AlSi_3O_8]$	6	Columnar, Plate	Red, Rose, Brown	White	Glass	2.6	Two groups of cleavage
17	Plagioclase	$Na[AlSi_3O_8]$~$Ca[Al_2Si_2O_8]$	6	Plate, Granular	White, Light yellow	White	Glass	2.7	Two groups of cleavage
18	Pyrite	FeS_2	6~6.5	Cube, Grainy, Block	Pale yellow copper	Green black	Metal	4.9~5.2	Uneven fracture
19	Peridot	$(Mg,Fe)_2[SiO_4]$	6.5~7	Granular	Olive green	White	Glass	3.3~3.5	Conchoidal fracture
20	Garnet	$(Ca,Mg)_3(Al,Fe)_2[SiO_4]_3$	6.5~7.5	Rhomboid dodecahedron, Granular	Variety	White	Glass, Grease	3~4	No cleavage
21	Quartz	SiO_2	7	Hexagonal biconical column, Block	Colorless, White	White	Glass, Grease	2.65	Conchoidal fracture
22	Andalusite	$Al_2[SiO_4]O$	1~2	Columnar, Radial	Light green and Light red	White	Glass	3.1~3.2	Two groups of cleavage

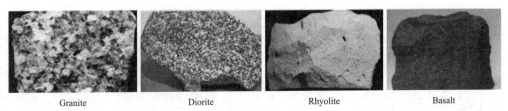

Granite Diorite Rhyolite Basalt

Fig. 2-18 Typical igneous rocks

Earth's crust.

Intrusive action means that magma deep underground rises along the fissures, but does not reach the surface. It condenses and crystallizes in a certain place below the ground to become rocks, which are called intrusive rocks. The rock formed by the condensation and crystallization of magma in a relatively deep part of the Earth's crust is called abyssal rock. The rock formed by condensation and crystallization when magma rises to a shallower part of the Earth's crust or close to the surface is called hypabyssal rock. Extrusive action, also known as volcanism, is the process of magma spilling over the surface until it condenses and becomes rock. Rocks formed by extrusive action are called extrusive rocks (or volcanic rocks). Therefore, according to the diagenetic environment and the tectonism, igneous rocks could be divided into abyssal rocks, hypabyssal rocks and extrusive rocks. Typical igneous rocks could be seen in Fig. 2-18.

(1) Chemical composition of igneous rock

The chemical composition of igneous rock is very complex, in which oxygen, silicon, aluminum, magnesium, iron, calcium, sodium, potassium, manganese, titanium, phosphorus, and hydrogen are the main elements.

(2) The mineral composition of igneous rocks

There are about more than 30 kinds of minerals that compose igneous rocks. According to the characteristics of the color and chemical composition, it can be divided into two types, i.e., light minerals and dark minerals.

1) Light minerals, including quartz, orthoclase, plagioclase, and muscovite. They are rich in silicon and aluminum; hence, they are also called aluminum-containing silicate minerals or silica-aluminum minerals.

2) Dark minerals, including biotite, pyroxene, hornblende, olivine etc. They are rich in iron and magnesium; hence, they are also called iron- and magnesium-rich silicate minerals.

However, not all of these minerals are present at the same time in a specific rock which is usually made up of only two or three major minerals. The minerals with more than 10% content in igneous rocks are called main minerals, which are the main basis for the rock classification. Gabbro, for example, is mainly composed of plagioclase and pyroxene. Granite is composed of quartz, feldspar (orthoclase), and mica (biotite) (Fig. 2-19). The relatively smaller amount of minerals in rocks, only 1% and 10%, are called secondary minerals which are the basis for further naming of rocks, but do not affect the division of major categories.

The mineral composition of igneous rocks can reflect both the chemical composition and the formation conditions of rocks. The chemical composition of magma is quite complex, but its high content of SiO_2 has the

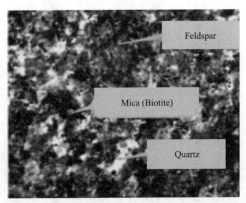

Fig. 2-19 Mineral composition of granite

greatest influence on the mineral composition of the rock. Therefore, according to silica (SiO$_2$) content and mineral composition, igneous rocks can also be divided into four categories: ultrabasic rocks (<45% silica), basic rocks (45%~52% silica), intermediate rocks (52%~65% silica), acid rocks (>65% silica). From ultrabasic rocks to acid rocks, their color and mineral composition change regularly with the increase of silica content.

(3) Texture of igneous rocks

The texture of igneous rocks refers to the degree of crystallization, grain size, morphology and the relationship among them. When the condensation is slow, the grain size is coarse and the crystal shape is intact. When the condensation is fast, many crystals buds will precipitate at the same time and compete with each other for growth space, resulting in fine grain and irregular crystal shape. When the rate of condensation is very fast, amorphous materials will be formed. It is one of the important marks to distinguish and identify igneous rocks, and it also directly affects the strength of rocks. The texture classification of igneous rocks is as follows.

1) Classification according to the degree of crystallization of minerals

Holocrystalline texture. Such rocks, composed entirely of crystalline minerals (Fig. 2-20 part A), commonly seen in abyssal and hypabyssal rocks such as granites and diorites.

Hemi-crystalline texture. Such rocks, composed of crystalline and amorphous minerals (Fig. 2-20 part B), are commonly found in hypabyssal and extrusive rocks such as rhyolite.

Amorphous texture. Such rocks are composed entirely of amorphous minerals, also known as glassy texture (Fig. 2-20 part C), which are characteristic of extrusive rocks such as obsidian.

2) Classification according to the absolute size of the mineral particles

Phanerocrystalline texture. The rocks are

A—Holocrystalline texture; B—Hemi-crystalline texture; C—Glassy texture

Fig. 2-20 Three textures according to the degree of crystallization

all composed of minerals with large crystalline grains that can be identified with the naked eye or a magnifying glass. According to the particle size of mineral particles, it can be divided into coarse grain texture (particle size >5 mm), medium grain texture (particle size 1~5 mm), fine particle texture (particle size 0.1~1 mm) and tiny particle texture (particle size <0.1 mm).

Semi-crystalline texture. The rocks are all composed of crystalline and tiny minerals. The crystal grains are invisible with the naked eye and magnifying glass, and can only be identified under a microscope.

3) Classification according to the relative size of the mineral particles

Equigranular texture (Fig. 2-21a). The minerals in the rock are all crystalline grains, and the crystalline grains of the same major minerals are roughly equal in size. The equigranular texture is a typical character of abyssal rock.

Inequigranular texture (Fig. 2-21b). For the main minerals in the rocks, their grain sizes are different, and in a continuous series. If the size of the two types of particles is very different (i.e., the difference is more than an order of magnitude), the larger ones with complete crystal forms are called phenocrysts, and slowly crystallize in the deep earth with higher temperature. The smaller ones are called stroma. When the stroma matrix is cryptocrystalline and vitreous, it is called porphyritic texture. If the matrix is crystalline, it is called porphyritic-like texture.

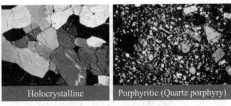

(a) Equigranular texture   (b) Inequigranular texture

Fig. 2-21 Typical textures of igneous rocks

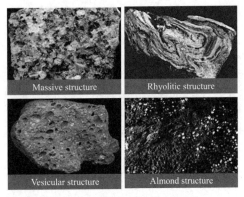

Fig. 2-22 Common structures of igneous rocks

The porphyritic texture is a unique texture of hypabyssal rocks and some extruded rocks. The reason for porphyritic structure formation is that phenocrysts are formed deep in the crust, and the matrix is formed after the phenocryst-containing magma rises to the shallow part of the crust or spits out to the ground. Porphyritic-like texture are mainly distributed in shallow intrusive rocks and some mesogenic and abyssal intrusive rocks. Phenocrysts and substrates of porphyritic-like textures are formed simultaneously in the same environment.

(4) Structure of igneous rocks

The structure of igneous rock refers to the appearance characteristics of the rock reflected by the shape, the size, the order and the filling method of minerals in the rock. The structure characteristics of igneous rocks are mainly determined by the environment in which the magma condenses. The typical structures of igneous rock are as follows (Fig. 2-22).

Massive structure. The minerals in the rocks are evenly distributed, without specific directional arrangement, and the rocks are uniformly dense blocks. It can be seen in most igneous rocks, in all intrusive rocks and in some extrusive rocks.

Rhyolite structure. It is one kind of structure in a rock consisting of colored stripes, elongated pores, and long bands of minerals arranged in a certain direction. It reflects the flow of magma after erupting to the ground, and can be found in extrusive rocks, such as rhyolite.

Vesicular structure. In the rapid condensation process of magma erupting to the ground, the gases or volatile substances contained in the magma tend to escape from the magma, therefore the pores of different sizes are formed in rock, which is called vesicular structure, commonly found in extrusive rock.

Aimond structure. The pores of vesicular structure rocks are filled with secondary minerals (i.e., calcite, quartz, etc.) to form an almond-like structure rock. Such structure is often found in extrusive rocks such as andesite.

(5) Classification of igneous rocks and common igneous rocks

1) Classification of igneous rocks

There are many kinds of igneous rocks in nature, which are different in mineral composition, texture and structure, attitude and genesis, so their engineering geological characteristics are also obviously different. The classification of igneous rocks is usually based on the chemical composition, mineral composition, texture and structure, formation conditions and attitude of rocks. First of all, according to the chemical composition of the igneous rock (i.e., mainly the content of SiO_2), they are divided into acid rocks, intermediate rocks, basic rocks and ultrabasic rocks. Secondly, according to the formation conditions of igneous rocks, they can be divided into extrusive rocks, hypabyssal rocks and abyssal rocks. On the basis of chemical composition and formation conditions, the attitude, texture and structure of igneous rocks should be taken into consideration. Consequently, the main types of igneous rocks classified accordingly are shown in Table 2-4.

Classification of igneous rocks

Table 2-4

Types of igneous rocks				Color →	Acid	Intermediate	Shallow → Deep	Basic	Ultrabasic
SiO$_2$ content					>65%	52%~65%		45%~52%	<45%
Genetic type				Main mineral	Quartz Orthoclase Plagioclase	Orthoclase Plagioclase	Hornblende Plagioclase	Plagioclase Pyroxene	Plagioclase Pyroxene
				Secondary mineral	Mica Hornblende	Hornblende Biotite Pyroxene Quartz<5%	Pyroxene Biotite Orthoclase<5% Quartz<5%	Peridot Hornblende Biotite	Hornblende Plagioclase Biotite
				Structure					
Extrusive rock		Lava flows	Attitude		Volcanic glass; Obsidian; Pumice etc.				
			Almond Vesicular Rhyolitic Massive	Amorphous (vitreous)					
				Cryptocrystalline porphyritic	Rhyolite	Cryptocrystalline porphyritic	Rhyolite	Cryptocrystalline porphyritic	Rhyolite
Intrusive rocks	Hypabyssal rock	Sill Dyke	Massive	Holocrystalline porphyritic Fine grain	Granite porphyry	Syenite porphyry	Diorite porphyrite	Diabase	Picrite porphyrite
	Abyssal rock	Stock Batholith		Crystallization porphyritic Coarse grain	Granite	Syenite	Diorite	Gabbro	Peridotite Pyroxenite

Fig. 2-23 Common igneous rocks

2) Common igneous rocks (Fig. 2-23)

① Granite. It belongs to abyssal rock, has holocrystalline equigranular texture and massive structure, and mostly is fleshy red, light gray, gray-white. The main mineral composition is quartz, orthoclase, plagioclase, and the secondary minerals are biotite and amphibole.

② Diorite. It is abyssal rock, has the holocrystalline equigranular texture and massive structure, and is gray, dark gray to black gray in color. The main minerals are plagioclase and amphibole, and the secondary are pyroxene and biotite.

③ Gabbro. It is abyssal rock, has the holocrystalline equigranular texture and massive structure, and is gray black and black in color. The main minerals are plagioclase and pyroxene, and the secondary are olivine, amphibole and biotite.

④ Diabase. It belongs to hypabyssal rock, and is usually grey-green or black-green, with special diabase texture (i.e., pyroxene is filled in the gap of plagioclase crystal frame).

⑤ Rhyolite. It belongs to extrusive rock, and is often gray, gray red, light yellow brown. Its mineral composition is the same as granite, with cryptocrystalline porphyry texture, and typical rhyolite structure. Its phenocryst is mainly quartz and orthoclase, and the matrix is usually vitreous.

⑥ Andesite. It belongs to extrusive rock, and is gray or hair purple, with porphyritic texture, has vesicular or almond structure, and its common phenocryst is plagioclase.

⑦ Basalt. It belongs to extrusive rock, and it characterized by cryptocrystalline fine grains or porphyritic texture, vesicular or almond structure, gray black to black in color. The main minerals are the same as that of gabbro.

⑧ Olivine rock. It belongs to abyssal rock with good crystallization. And it is dark green, and has granular texture and massive structure. Olivine belongs to orthorhombic system, is olive and yellow green with glass luster, hardness of 6.5~7, shell fracture.

2.3.2 Sedimentary Rock

Sedimentary rocks are layered rocks formed by a series of geological processes such as transportation, sedimentation and diagenesis under the normal temperature and pressure on the surface of the Earth's crust.

The formation of sedimentary rocks is a long and complicated geological process. All kinds of rocks exposed on the surface will gradually decompose and break, or become rock fragments, or become fine-grained clay minerals, or become other dissolved substances after long-term weathering and destruction. These little bits of our earth are washed downstream where they settle to the low-lying places of the rivers, lakes, and oceans to form loose sediments. Layer after layer of eroded earth is deposited on top of earth. Those sediments will gradually become sedimentary rocks after actions including compaction, cementation, and re-crystallization.

(1) Material composition of sedimentary rocks

The material composition of sediments includes the following categories.

Clastic material. Clastic material is composed mainly by the weathering products of the original rock, partly by the residual clastic from the destructed original rock, and the clastic mineral with strong weathering resistance such as quartz, feldspar and muscovite.

Clay minerals. They are secondary minerals generated by weathering and decomposing the original rock, such as kaolinite, montmorillonite, and hydromica.

Chemically deposited minerals. Chemically deposited minerals are minerals formed by chemical or biochemical deposition, such as calcite, dolomite, gypsum, halite, oxides and hydroxides of iron and manganese.

Organic minerals and biological residues. They are organic minerals formed by biological remains or organic chemical changes, such as shells, peat and other organic matters.

(2) Texture of sedimentary rock

The texture of sedimentary rock refers to the state, size and their connection forms of minerals. There are generally four kinds of textures in sedimentary rock, i.e., clastic texture, pelitic/clay texture, crystalline texture, biological texture.

Clastic texture is a texture consisting of more than 50% of the clastic material greater than 0.005 mm in diameter as shown in Fig. 2-24 (a). Generally, according to the size of clastic particle size, it can be divided into three types First is gravel texture (particle size>2 mm). Second is sand texture, which includes coarse grained texture (particle size 0.5~2 mm), such as coarse-grained sandstone; medium grained texture (particle size 0.25~0.5 mm), such as medium grained sandstone; fine grained texture (particle size 0.05~0.25 mm), such as fine-grained sandstone. Third is silty texture (particle size 0.005~0.05 mm), such as siltstone.

Pelitic/clay texture refers to the structure composed of more than 50% of fine debris, and its clay mineral particles are less than 0.005 mm in diameter, such as mudstone, shale.

Biological texture refers to a texture containing over 30% of the remains or fragments of living things, such as coral texture and shell texture formed, as shown in Fig. 2-24 (b).

(a) Clastic texture　　(b) Biological texture

Fig. 2-24 Texture of sedimentary rocks

(3) Structure of sedimentary rocks

The structure of sedimentary rock refers to the spatial distribution and their arrangement relationship of the constituent parts of the rock. The most important structures of sedimentary rocks are bedding, bedding planes and fossils.

1) Bedding. It refers to the arrangement of particles in distinct layers of different deposited particles in size, shape, color and composition due to the change of sedimentary environment during the formation of sedimentary rocks, as shown in Fig. 2-25.

According to the thickness of the rock layer, it can be divided into huge thick layer (>1 m), thick layer (0.5~1 m), medium thickness layer

(a) The Wave bedding (b) Cross-bedding

Fig. 2-25 The Wave (USA) and red sand-mudstone bedding geomorphologic landscape (Chongqing, China)

(0.1~0.5 m) and thin layer (<0.1 m). Thin layers sandwiched between thick layers are called intercalations. Some rock layers are thicker at one end, and being gradually thinner at the other end to disappear, such phenomenon called pinch-out layer. If both ends are pinched out in a small distance a thicker middle layer, it is called lenticle (Fig. 2-26d).

Bedding structure can be divided into the following four types according to the morphology of bedding (Fig. 2-26).

Horizontal bedding refers to a bedding consisting of a series of thin layers that are straight and parallel to the bedding (Fig. 2-26a). It usually occurs in fine-grained rocks (clays, fine silts, micrite, etc.). It is formed through slow deposition of suspended matter or solutions powered by hydrodynamics (such as river bank zone, closed bay, deep water zone of sea and lake).

Monocline bedding consists of a series of fine layers oblique crossing the bedding, the bedding of which inclines in the same direction and is parallel to each other (Fig. 2-26b). It intersects diagonally with the upper and lower planes, which are parallel to each other. Monocline bedding, often seen in the sediment of a riverbed or coastal delta, is caused by one-way current.

Cross bedding is formed by overlapping of inclined bedding with different directions (Fig. 2-26c), which is caused by frequent changes of water flow and is usually found in fluvial sediments.

Wavy bedding is undulating and its general direction is roughly parallel to the plane. Such bedding can be divided into parallel undulating bedding and oblique wavy bedding (Fig. 2-26d). Wavy bedding is formed in the case of fluid fluctuation.

The thinning, pinching and lensing of the rock can change its strength and water permeability in different directions. Soft weak interlayer is easy to cause bedding slip of overlying strata. Sedimentary rocks tend to split along bedding planes.

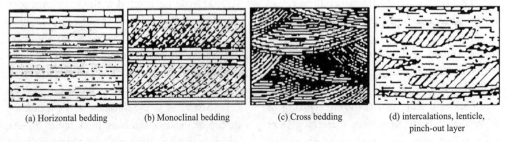

(a) Horizontal bedding (b) Monoclinal bedding (c) Cross bedding (d) intercalations, lenticle, pinch-out layer

Fig. 2-26 Bedding types of sedimentary rocks

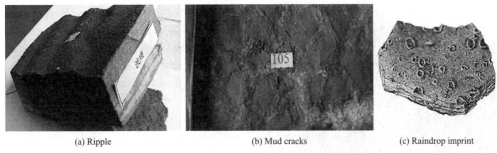

(a) Ripple (b) Mud cracks (c) Raindrop imprint

Fig. 2-27 Bedding structure (ripple, mud cracks and raindrop imprint)

2) Bedding plane refers to the traces left by water flow, wind, biological activities, sun exposure, etc., such as ripple, mud cracks, and raindrop imprint, as shown in Fig. 2-27. Ripple refers to the undulating traces formed on the surface of sedimentary layers due to the action of wind, water or wave, which can indicate the movement of water or wind from gentle slope to steep slope when ripple is forming. Mud cracks refer to the polygonal mesh cracks due to water loss caused by sun exposure when the sediments that are exposed to water surface before they are consolidated and contracted.

Fig. 2-28 Fossils (Wansheng Stone Forest, Chongqing, China)

3) Fossils are the remains of plants and animals preserved in sedimentary rocks by petrification, such as trilobites, scales, and clam shells. The formation environment and geological age of the rock can be determined according to fossils, as shown in Fig. 2-28.

4) Concretion is a block mass that differs in composition, texture, structure, and color from surrounding sediments. There are many concretion forms, such as spherical, ellipsoidal, and irregular mass. For example, chert concretion commonly found in limestone (Fig. 2-29 and Fig. 2-30), are mainly formed by colloidal condensation of SiO_2 during deposition. The calcareous concretion in the loess is formed by the re-deposition of $CaCO_3$ dissolved by groundwater from the sediments.

Fig. 2-29 Concretions in limestone

Fig. 2-30 Phosphate concretions

5) Stylolite, common in limestone and dolomite (Fig. 2-31), refers to the jagged undulation curve in the rock section. It is formed in the diagenetic period when materials experience pressure dissolution under the pressure of the overlying rock layer. Calcite and dolomite are dissolved and taken away by acidic solution, and quartz by

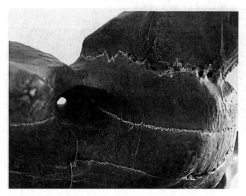

Fig. 2-31 Stylolite

alkaline solution along both sides of the rock surface. Along with the uneven entrainment of some components along the vertical pressure direction, serrated undulating sutures are formed. It is commonly found in limestone and dolomite.

The bedding, bedding plane and the presence of fossils of sedimentary rocks are the important characteristics that distinguish sedimentary rocks from igneous rocks in terms of structure.

(4) Classification of sedimentary rocks and common sedimentary rocks

1) Classification of sedimentary rocks

Sedimentary rocks can be divided into the following types according to their constituent material composition and texture characteristics (Table 2-5).

2) Common sedimentary rocks

Fig. 2-32 shows the typical sedimentary rocks in nature and the characteristics of main sedimentary rocks.

Conglomerate and breccia, with gravel texture, are composed of more than 50% of coarse clastic cements greater than 2 mm. Those cemented by rounded gravel are called conglomerates. Those cemented by angular breccia are called breccia. The lithological components of breccia are relatively single, while the lithological composition of conglomerate is generally complex, consisting of a variety of rock fragments and mineral grains. The components of cements are calcareous, argillaceous, ferric and siliceous.

Sandstone, with sand-like texture, is composed of more than 50% of sand particles with diameters between 0.05 and 2 mm, and less than 25% of clay. According to the mineral composition of sand particles, sandstone can be divided into quartz sandstone, arkose and litharenite. According to the size of sand particle size, it can be divided into coarse sandstone, medium sandstone and fine sandstone. According to the composition of cements, sandstone can be divided into siliceous sandstone, ferric sandstone, calcareous sandstone and argillaceous sandstone. Siliceous sandstone is light in color, high in strength with good resistance to weathering. Mudstone is generally yellow-brown, and easy to be softened, with poor strength and stability. Ferric sandstone is often fuchsia or brownish red. Calcareous sandstone is white or gray, whose strength and stability is between siliceous and argillaceous sandstone.

Siltstone, with silty texture, is often with clear horizontal bedding. It is made of more than 50% of silt with particle size between 0.005 mm and 0.05 mm, and less than 25% of clay.

Shale, formed by the dehydration and cementation of clay, contains mainly clay minerals. Most shale has obvious thin bedding and is laminated. Shale includes siliceous shale, earthy shale, sandy shale, calcareous shale and carbonaceous shale.

Mudstone is similar in composition to shale, often in thick layers. With kaolinite as the main component of the mudstone, it is often gray or yellow white, with strong water absorption ability, making it easy to be softened with water. Mudstone with microcrystalline kaolinite as its main component, is often white, rose or light green, with slippery surface, little plasticity, and strong water absorption ability which makes it expands rapidly after absorbing water.

The mineral composition of limestone is mainly calcite, and a small amount of dolomite and clay minerals. It is often dark gray or light

Classification of sedimentary rocks   Table 2-5

Rock types			Texture	Classification and definition of rocks	Major subclasses and their constituent substances
Clastic rock	Volcanic clastic rock		Agglomeratic texture (particle size >100 mm)	Volcanic agglomerates	There are mainly more than 100 mm of lava fragments, volcanic dust and other compaction cementation
			Breccia texture (particle size >2~100 mm)	Volcanic breccia	Mainly composed of 2 ~ 100 mm lava debris
			Tuffaceous texture (particle size <2 mm)	Tuff	It is composed of more than 50% volcanic ash with particle size less than 2 mm
	Sedimentary clastic rock	Clastic texture	Gravel texture (particle size >2 mm)	Conglomerate	Breccia: formed by cementing angular fragments Conglomerate: composed of rounded gravel that has been cemented together
			Sandy texture (particle size 0.05~2 mm)	Sandstone	Quartz sandstone: quartz (>90%), arkose (<10%) Feldspar: quartz (<75%), feldspar (>25%), detritus (<10%) Litharenite: quartz (<75%), feldspar (<10%), detritus (>25%)
			Silt texture (particle size 0.005~0.05 mm)	Siltstone	Mainly composed of quartz, feldspar powder, clay and clay minerals
	Clay rock		Pelitic/clay texture (particle size <0.005 mm)	Mudstone	Mainly composed of clay minerals such as kaolinite, microcrystalline kaolinite and hydromica
				Shale	Clayey shale: composed of clay minerals Carbonaceous shale: composed of clay minerals and organic matter
Chemical and organic chemical rock			Crystalline and biological texture	Limestone	Limestone: calcite (>90%), clay minerals (<10%) Marl: calcite (50% ~ 75%), clay minerals (25% ~ 50%)
				Dolomite	Dolomite. dolomite (90% ~ 100%), calcite (<10%) Calcite. dolomite (50% ~ 75%), calcite (25%~50%)

Fig. 2-32 Common sedimentary rocks

gray. The crystalline limestone formed by pure chemical action has crystalline texture with fine grains however. Crystalline limestone with obvious grains can be formed by recrystallization. Limestones formed by biochemical processes are often rich in organic remains.

The mineral composition of dolomite is mainly dolomite, and some calcite and clay minerals, with crystalline texture. Pure dolomite is white, and it could appear different colors with different impurities. Its features are similar to that of limestone, but its strength and stability is higher than limestone, therefore it is a good choice for buildings.

Gypsum is made up of sulfate mineral and formed by sea water evaporating in massive prehistoric basins. It is very soft and used to make casts, molds, and wallboards.

2.3.3 Metamorphic Rock

The crustal movement and magmatic activities can cause changes in the physical and chemical environment. The mineral composition, texture and structure of the preformed rocks (magmatic rocks, sedimentary rocks and early metamorphic rocks) tend to change in the solid state under the action of high temperature, high pressure and other chemical factors. The new rock formed by the change is called metamorphic rock.

Metamorphism occurs essentially in the original position of the original rock while it remains in solid state. Metamorphic rocks formed by magmatic rocks are called positive metamorphic rocks. Metamorphic rocks formed from sedimentary rocks are called parametamorphic rocks. The attitude of positive metamorphic rocks retains the attitude of proto-magmatic rocks, while the attitude of para-metamorphic rocks retains the attitude of sedimentary rocks.

(1) Mineral composition of metamorphic rocks

The mineral composition of metamorphic rocks is very complex, and can be divided into two categories. One type is common in igneous rocks or sedimentary rocks. The other is the metamorphic rocks produced in the metamorphic minerals, such as graphite, talc, serpentine, garnet, chlorite, sericite, wollastonite, disthene, andalusite, known as metamorphic minerals.

(2) Texture of metamorphic rock

The main texture of metamorphic rocks is shown as follows.

1) Palimpsest texture. In the process of metamorphism of the original rock, due to the incomplete recrystallization and metamorphic crystallization, the mineral composition and structural characteristics of the original rock are partially retained, which is called the palimpsest

texture. For example, after the metamorphism of argillaceous sandstones, the argillaceous cements metamorphize into sericite and chlorite, and the clastic minerals such as quartz do not change and retain, forming blasto-psammitic texture. Others include blasto-porphyritic texture, blasto-psephitic texture, blasto-granitic texture (Fig. 2-33a), blasto-pelitic texture and so on.

2) Crystalloblastic texture. The texture formed by recrystallization, metamorphic crystallization or recombination of rock in solid state is called crystalloblastic texture. This is the most common texture in metamorphic rocks. The minerals in this type of structure are usually arranged in a specific directional.

According to the relative size of metamorphic mineral particles, crystalloblastic texture can be divided into equal-grain metamorphic texture, unequal-grain metamorphic texture and porphyritic metamorphic texture (Fig. 2-33b); According to the absolute size of metamorphic mineral particles, such texture can be divided into coarse-grained metamorphic texture (main mineral particles diameter>3 mm), medium-grain metamorphic texture (1~3 mm), fine-grain metamorphic texture (0.1~1 mm), microscopic metamorphic texture (<0.1 mm). According to the shape of metamorphic mineral particles, the texture can be divided into granoblastic texture, scale-like metamorphic texture and fibroblastic texture.

3) Crush texture. Rocks tends to suffer directional pressure at low temperature. Once the pressure exceeds their strength limit, rocks would break and displace, becoming fragments or even powder which will be cemented together to form crush texture. It is a common texture in dynamic metamorphic rocks. According to the degree of fragmentation, it can be divided into cataclastic texture (Fig. 2-33c), porphyroclastic texture, mylonitic texture, etc.

(3) Structure of metamorphic rocks

The structure of metamorphic rocks refers to the spatial distribution and arrangement of various minerals in them. After metamorphism, rocks often form some new structural features, most important characteristics of metamorphic rocks, which are unique to be differentiated from the other two types of rocks. The main structures of metamorphic rocks include schistosity structure, massive structure and mixture structure. As for schistosity structure, it mainly contains slate structure, phyllitic structure, schistose structure, and gneissic structure, as shown in Fig. 2-34.

Metamorphic rock with slate structure has a parallel, dense and flat fracture plane, called cleavage plane, along which the rock can be easily broken into plates. Such rocks often have a blastopelitic structure. The protolith is basically not recrystallized, and the mineral grains in the rock are too small to be distinguished by naked eyes, with only a small amount of sericite or chlorite, as shown in Fig. 2-34. It is formed when the rock is subjected to light directional pressure.

Metamorphic rock with phyllitic structure

(a) Blastogranitic texture (b) Porphyritic metamorphic texture (c) Cataclastic texture

Fig. 2-33 Textures of metamorphic rocks

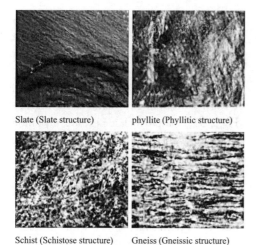

Slate (Slate structure)   phyllite (Phyllitic structure)
Schist (Schistose structure)   Gneiss (Gneissic structure)

Fig. 2-34 Tectonics of metamorphic rocks

is often in the form of thin plates, in which the components have been basically recrystallized and arranged in a directional way. However, the degree of crystallization is so low that minerals cannot be distinguished by naked eyes. There is a strong silk luster on the natural fracture surface of the rock, which is caused by sericite and chlorite. Metamorphic rocks with phyllitic structure can be easily broken into sheets along the foliation plane, and there are often small wrinkles on the foliation plane, as shown in Fig. 2-34.

Metamorphic rock with schistose structure has a large number of sheet and columnar minerals such as mica, amphibole and chlorite, which are arranged in parallel orientation under the long-term action of directional extrusion stress. The components of the rock are all recrystallized and the mineral grains are visible to the naked eyes. Rocks with such structure are obviously anisotropic and prone to crack along the foliation plane, as shown in Fig. 2-34. Their strength, water permeability and weathering resistance also change with the direction.

Metamorphic rock with gneissic structure is mainly composed of granular minerals such as quartz and feldspar, with flake and columnar crystalline minerals distributed in roughly parallel intermittent bands. The degree of crystallization is relatively high and is a common structure in gneiss (Fig. 2-34).

Phyllitic, schistose and gneisses are directional structure, which make metamorphic rocks tend to split into irregular sheets or lentils, which is also known as fragmentation.

(4) Classification and common metamorphic rocks

1) Classification of metamorphic rocks

Regional metamorphic rocks are mainly classified according to their structure, while the massive metamorphic rocks according to their mineral composition, and the dynamic metamorphic rocks according to the texture reflecting the degree of fragmentation. Their classification is shown in Table 2-6.

When identifying metamorphic rocks, the structure of rocks should first be observed. According to the structure, metamorphic rocks can be divided into two types, i.e., schist structure and massive structure. The classification of the rocks can then be further determined by analyzing the subgroups to which the rocks belong according to the foliation characteristics, structure and the main mineral composition.

2) Common metamorphic rocks

Fig. 2-35 gives the typical metamorphic rocks in nature. The characteristics of main metamorphic rocks are shown as follows.

Slate tends to be dark gray to black, the mineral particles of which are very fine, and mainly include sericite, clay minerals and chlorite. It mainly has palimpsest texture or cryptoblastic texture, and plate structure. This kind of rock is very dense, easy to crack into thin plates with uniform thickness. It usually has a brittle sound with hammer strikes which is a feature to distinguish it from shale.

Phyllite tends to be gray, green to black, is mainly composed of cryptocrystalline sericite, chlorite, with palimpsest or crystalloblastic texture, and phyllosic structure, and a strong silk luster on its surface.

Schist is mainly composed of mica, chlorite, talc and other lamellar minerals, with crystalloblastic texture and schistose structure. It is easy to crack into sheets along the lamellar

Classification of metamorphic rocks Table 2-6

Rock types	Rock name	Structure	Texture	Main mineral composition	Metamorphic type
Foliated rocks	Slate	Slate	Palimpsest texture / Partial crystalloblastic texture	Clay minerals, mica, chlorite, quartz, feldspar, etc.	Regional metamorphism (increasing from slate to gneiss)
	Phyllite	Phyllite	Microscale crystalloblastic texture	Sericite, quartz, feldspar, chlorite, calcite, etc.	
	Schist	Schist	Exocrystalline scaly crystalloblastic texture	Mica, amphibole, chlorite, graphite, talc, garnet, etc.	
	Gneiss	Gneiss	Granoblastic texture	Quartz, feldspar, mica, amphibole, pyroxene, etc.	
The massive rocks	Marble	Massive	Granoblastic texture	Calcite, dolomite	Contact metamorphism or regional metamorphism
	Quartzite		Granoblastic texture	Quartz	
	Skarn		Anisomorphic texture	Garnet, pyroxene, wollastonite (calcareous skarn)	Contact metamorphism
The massive rocks	Serpentine		Aphanitic texture	Serpentine	Metasomatism
	Greisen		Granoblastic texture	Muscovite, quartz	
Tectonically fractured rocks	Fault breccia		Cataclastic texture / Brecciated texture	Rock debris, mineral debris	Dynamic metamorphism
	Mylonite		Mylonite texture	Feldspar, quartz, sericite, chlorite	

Fig. 2-35 Common metamorphic rocks

surface.

Gneiss is mainly composed of feldspar and quartz, as well as a small amount of metamorphic minerals such as biotite, amphibole and garnet. Its mineral crystals are coarse and are banded-distributed. It has granoblastic texture with typical gneissic structure.

Marble is white or gray, and mainly composed of calcite and dolomite, with crystalloblastic texture and massive structure. Marble is formed by recrystallization of limestone or dolomite. The marble (Dali Stone in Chinese) derives its name from Dali City, Yunnan Province, China, where is abound in marble. White marble and marble with various patterns are commonly used as building materials and decorative stones.

Quartzite is composed of sandstone that has been metamorphosed. It forms from sandstone contacted with deeply buried magmas. Quartzite looks similar to its parent rock. The best way to tell quartzite from sandstone is to break the rocks.

Serpentine is mainly composed of serpentine. The olivine and pyroxene in the original rock are serpentinized by the metasomatism of medium and low temperature hydrothermal fluids or regional metamorphism of the ultrabasic rocks. The rock is generally yellow-green to black-green, with compact block and low hardness. The weathered surface is usually grayish-white, with sometimes reticulated structure.

Mylonite is one kind of fine-grained dynamic metamorphic rock formed after the original rock is severely crushed. Under microscope, mylonite is mainly composed of fine quartz, feldspar and a few new recrystallized minerals (sericite, chlorite, etc.). The grain size of mineral fragments is generally less than 0.5 mm. Sometimes a small amount of coarser proto-rock fragments with augen porphyroclast could be seen, which are arranged in obvious orientation, forming a mylonitic texture. Due to the difference in the degree of crushing or the composition of the crushed material, striped structure can be formed. The lithology of mylonite is hard and compact, and it looks similar to siliceous rock with the naked eyes, and is often found in fault fracture zones.

Cataclasite is a dynamic metamorphic rock and usually occurs in fault zones. The differences between cataclasite and fault breccia are that first cataclasite is more finely broken. And second the crushing of mineral grains in the original rock is more severe.

2.3.4 Comparison of Geological Characteristics of Different Rocks

(1) Comparison of texture features

Igneous rocks are formed directly from the condensation of molten magma at high temperature and have obvious crystalline structure, which is reflected in the sequence of condensation and crystallization of the combined minerals.

Sedimentary rock is formed by weathering, denudation, transportation, deposition, compaction and cementation of original rocks, and has obvious structural characteristics of material deposition, that is, clastic texture, pelitic texture and biological texture. Its crystalline texture reflects the chemical properties of precipitation or recrystallization in solution.

Metamorphic rocks are formed from original rocks affected by different degrees of metamorphic factors. In terms of texture, it has both inheritance and uniqueness, and presents crystalloblastic texture, palimpsest texture and crush texture. The texture of metamorphic rocks reflects the directivity of various minerals to recrystallize under directional pressure in solid state.

(2) Comparison of structure features

Igneous rocks show different structures with different magmatic properties, conditions and movement states during solidification. During the formation of intrusive rocks, different mineral crystals aggregate into blocks due to the cohesion among mineral crystals during the cooling of magma. Extrusive rocks often form rhyolitic, vesicular, almond and dense massive structures due to minerals are vitreous or cryptocrystalline.

The lithology differences of sedimentary rocks are from factors including the nature of external dynamic geological action, palaeogeographic environment, material source and sedimentary conditions. But on the macroscopic scale, they all have layered texture and bedding structure, among which the thick layer of rock precipitated by biochemistry can be regarded as massive structure.

Metamorphic rocks show various structures with the variation of environment, mode and intensity of metamorphism of the original rock. However, the most common one is schistosity structure. It is a remarkable structural feature of metamorphic rocks, which is distinguishable from igneous rocks and sedimentary rocks. In addition, there is also massive structure with homogeneous recrystallization. The main differences between the three types of rocks are summarized in Table 2-7.

Geological characteristics of the three types of rocks　　　　Table 2-7

Rock types	Cause	Composition	Texture	Structure	Representative rock
Igneous rock	High temperature High pressure magma cooling	Silicate Metallic sulfide	Porphyritic texture Holocrystline texture	Massive structure Rhyolitic structure Vesicular structure	Granite Rhyolite Basalt
Sedimentary rock	Atmospheric pressure Rock weathering Compaction cementation	Detrital material Clay minerals Chemicals Organic matter	Clastic texture Clay texture Crystalline texture Biological texture	Bedding structure Bedding plane structure	Sandstone Mudstone Limestone
Metamorphic rock	Metamorphic rock	Metamorphic minerals	Crush texture Crystalloblastic texture Palimpsest texture	Schistose structure Gneissic structure Phyllitic structure Massive structure	Schist Gneiss Phyllite Marble

Quiz

1. What are minerals? What are the morphological characteristics of minerals?

2. What are the physical properties of minerals?

3. Comparison of characteristics between granite and marble, quartzite and marble.

4. How to distinguish sedimentary, metamorphic and igneous rocks?

5. What is the main texture and structural features of sedimentary rocks?

6. What is the main texture and structural features of metamorphic rocks?

7. What is the main texture and structural features of igneous rocks?

8. What is the hardness of minerals? How to divide the hardness of minerals?

9. What are the differences between bedding and foliation?

10. How to understand the interconversion between the three groups of rocks?

Chapter 3
Geological Structures

3.1 Geological Process

During the development of geological history, the effects causing the continuous changes and development of the composition, internal structure and surface morphology of the Earth's crust are called geological action. Engineering geology divides geological processes into two categories: natural geological process and engineering geological process (man-made geological action). Natural geological action can be divided into endogenic geological process and exogenic geological process according to their dynamic sources.

Endogenic geological process refers to the geological action caused by the internal energy of the Earth, such as the Earth's rotation energy, gravitational energy and thermal energy generated by the transformation of radioactive elements, including tectonic movement, magmatism, metamorphism and seismic action.

Exogenic geological process refers to the geological action caused by the energy outside earth, which mainly comes from the solar radiation heat in the universe and the gravitational action of the moon. According to geological forces, exogenic geological process includes weathering, denudation, transportation, sedimentation and consolidation diagenesis.

Engineering geology process refers to the geological effects caused by human activities, for example, surface deformation, collapse and landslide caused by the exploitation of underground resources; land inundation, salinization, swamping caused by construction of water conservancy projects; reservoir bank landslide; reservoir induced earthquake, etc.

The manifestation of geological action is listed in Table 3-1.

The manifestation of geological action   Table 3-1

Type		Definition
Endogenic geological process	Tectonic movement	The mechanical movement of crustal rocks including deformation and displacement caused by the Earth's internal dynamics (bending, dislocation, etc.). Horizontal movement refers to the movement of crustal or lithospheric blocks in the horizontal direction, such as the separation of adjacent blocks, shearing and stagger off. Vertical movement refers to the differential rise or fall of adjacent blocks of the crust or lithosphere or different parts of the same block
	Magmatic action	Under the influence of crustal movement, the magma inside the Earth will move in the direction of decreasing external pressure, rise and intrude into the crust or eject out of the ground to cool down and solidify into rock. This process is called magmatism
	Earthquake action	The rapid tremor in a certain range of the Earth's crust caused by the sudden release of the long-term accumulation of energy inside the earth that has reaches a certain limit because of the crustal movement. According to the causes of earthquakes, earthquake action can be divided into tectonic earthquakes, volcanic earthquakes, subsidence earthquakes, and induced earthquakes
	Metamorphism	The physical and chemical conditions have changed due to crustal movement and magmatism. The effect of the rock changing its composition and structure is called metamorphism

Continued

Type		Definition
Exogenic geological process	Weathering	Weathering is a damage that causes the rocks that make up the surface of the Earth's crust to break and in-situ decompose, under the comprehensive influence of temperature changes, gas, water and biological factors. It can be divided into physical, chemical, and biological weathering
	Denudation	Denudation is the stripping of matters damaged by weathering damage from the original place
	Transportation	The weathered and eroded rocks are transported to other places by flowing water, wind, glaciers or other media
	Sedimentation	The material to be transported, due to the weakening of the transport capacity of the transport medium, changes in the physical and chemical conditions of the transport medium or the process of separation from the transport medium due to biological action and the formation of sediments is called sedimentation
	Consolidation cemaction	Consolidation cemaction refers to the process of loose deposits consolidating into a hard rock due to physical and chemical processes such as compaction, cementation and recrystallization because of the increase of pressure, temperature and effects from certain chemical solutions
Anthropogenic geology action	Engineering geology action	The changes of original geological conditions because of the process of human engineering activities

3.2 Geological Time and Geochronological Scale

3.2.1 Geological Time and its Determination Methods

Geological time is a method of ordering and measuring past events. The ordering of events usually uses a group of observational methods, which are known as absolute dating methods and relative dating methods.

Absolute geological age refers to the actual number of years since the formation of the stratum, and it is expressed in how many years ago. The relative geological age refers to the sequence of the strata formation and the relatively new-old relationship of the strata, which is determined by the relationship between the relative strata of the lithostratigraphic unit and the known adjacent lithostratigraphic units. The absolute geological age can explain the exact time of the strata formation, but it cannot reflect the geological process of the strata formation. The relative geological age does not include the time concept represented by "year", but it can explain the sequence of strata formation and the relative relationship between new and old strata. In engineering geological work, the relative geological age is more frequently used.

(1) Absolute Dating methods

Absolute dating methods calculate the real ages with year as the unit. There are two principal methods: One relies on event with very strong annual cyclicity in the geological record, such as tree growth rings, and coral growth cycles varves (annual clay sediment layers). The other method relies on the decay of naturally occurring radiogenic isotopes, which is more commonly used nowadays.

Radiometric dating is not without difficulties, but is by far the most widespread method applied to determine absolute ages between 100 and 4500 Ma.

The principle of such method is based on the fact that all radioactive elements have a fixed decay constant (λ), the parent isotope content (N) and the child isotope content (D) left after the radioactive isotope decay in the mineral can be detected. The age of a certain radioisotope can be calculated according to the following formula (t):

$$t = \frac{1}{\lambda} \ln \left(1 + \frac{D}{N}\right) \quad (3\text{-}1)$$

(2) Relative dating methods

Relative dating methods use geological principles to place events in chronological order, which include four methods: strata superposition method, fossil succession method, the method of rock layers contact relationship and lithology comparison method.

1) Strata superposition method

The strata superposition method is the basic method to determine the relative geological time.

The principle of superposition states that older beds are covered by younger beds (Fig. 3-1 and Fig. 3-2). In other word, the youngest bed is usually at the top in a sedimentary sequence. When the rock layer is tilted due to tectonic movement but not inverted, the rock layer above the inclined plane is new and the rock layer below the inclined plane is old. When the tectonic movement reverses the sequence of the rock strata, which is called strata inversion, the old rock strata will be overlaid on the new rock strata (Fig. 3-3).

2) Fossil succession method

According to the law of biological evolution, from ancient times to the present,

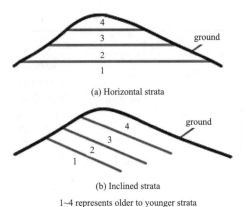

(a) Horizontal strata

(b) Inclined strata

1~4 represents older to younger strata

Fig. 3-1 Determination of the relative geological time (when the strata orders are regular)

Fig. 3-2 Stratigraphic sequence of Jiwei Mountain, Wulong, Chongqing, China

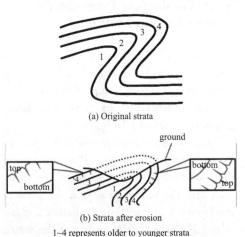

(a) Original strata

(b) Strata after erosion

1~4 represents older to younger strata

Fig. 3-3 Determination of the relative geological time (when the strata orders are reverse)

biology has always developed from lower to higher, from simple to complex organism. Therefore, the rock layers deposited in different geological ages will contain paleontological fossils with different characteristics. Rock layers containing the same fossils, no matter how far apart, were formed in the same geological time (Fig. 3-4). Therefore, as long as the geological times of the standardized stones contained in the rock formations are determined, the geological times of these rock formations will be determined accordingly.

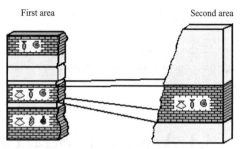

Fig. 3-4 The principle of fossil succession states

3) The method of rock layers contact relationship

The method of rock layers contacts relation states that the contact relations among rock layers are the history records of crust movement, magma activities (Table 3-2).

4) Lithology comparison method

Rocks formed in the same period and under the same geological environment have the same color, composition, structure and other lithological characteristics and sequence laws. Therefore, the age of the rock strata in a certain area can be determined according to the comparison of lithological characteristics.

Stratigraphic contact   Table 3-2

Contact of igneous rock and sedimentary rock	Intrusive contact	If the igneous rock mass is interspersed in the sedimentary rock after the formation of the sedimentary rock, the age of the sedimentary rock is older than that of the igneous rock
	Sedimentary contact	After the igneous rock is formed, the upper part is covered by sedimentary rock after weathering and denudation. The sedimentary rock is newer than the igneous rock
Igneous rock and igneous rock contact		When igneous rock masses intersect each other, the interspersed rock mass is older
Sedimentary rock and sedimentary rock contact	Conformable contact	Accumulate sequentially, the new stratum is upper and the old is lower
	Parallel unconformable contacts	Due to the ups and downs of the Earth's crust, there are sedimentary discontinuities in the middle, and part of the stratum is missing, but the upper and lower sets of rock formations have the same attitudes
	Angular unconformable contact	Some strata are missing between the adjacent new and old strata, and their attitudes are different from each other, and layers intersect at certain angle. There are obvious signs of weathering and erosion on the denuded surface, often with bottom conglomerate

3.2.2 Geochronological Scale

In the geological history, each geological age has a corresponding stratigraphic formation, which is called a chronostratigraphic unit. The chronostratigraphic units corresponding to the various eons, eras, periods, epochs, and period are eonothem, erathem, system, series, and stage.

3.3 Attitude of Strata and Contact Relationships

3.3.1 Attitude of Strata

The stratum: The layered rocks composed of the same lithology and are limited by two parallel or approximately parallel planes.

Attitude of stratum: The space location and state of stratum. Including three indexes: strike, dip, and dip angle.

(1) Strike

The extending direction of stratum plane (see the straight line AOB in Fig. 3-5).

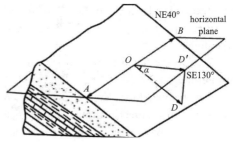

Fig. 3-5 Attitude elements of strata

Strike line: The intersection of the stratum plane and any horizontal plane, it has two directions whose difference is 180°.

(2) Dip

The direction of projection line of maximum inclination line onto the horizontal plane (see the ray OD' in Fig. 3-5). Dip and strike are at right angles to each other.

(3) Dip angle

The acute angle between stratum plane and horizontal plane (see the angle α in Fig. 3-5). Generally, the angle between the maximum inclination line and dip line, which is called true dip angle.

Fig. 3-6 Attitude indexes and their measurement methods of strata

3.3.2 Determination and Expression Methods of Strata Attitude

The geological compass can be used to directly measure the attitude elements of the strata exposed on the surface (Fig. 3-6). Since the strike of the rock formation is perpendicular to the dip, the strike of the rock is generally not measured directly, but after obtaining the dip azimuth, add or subtract 90° to obtain the strike azimuth.

Specific measurement methods are as follows.

(1) Strike

The long side of compass should be attached onto the strata plane, then we turn the compass until the bubble is at the center of bottom leveling instrument. The scale value which arrow points to is the strike. Since strike is a line with two extending directions, so both of two pointers can give right strike direction. i.e., NE30° and SW210, which mean one straight line.

(2) Dip

Put north end of compass pointing to the strata inclined direction, south end of compass attaches onto the strata plane, then turn the compass until the bubble is at the center of bottom leveling instrument. The scale value which white pointer points to is the dip.

If it is difficult to find the suitable location, then attach south end of compass onto the strata bottom plane, and read black pointer value. If it's difficult to read black pointer when measuring the bottom surface, attach the south end of the compass to the strata bottom plane, and read another pointer.

(3) Dip angle

Let compass stand on the strata plane, and attach long side of the compass to the maximum inclined line, move the compass side to side, and turn the wrench at the back of compass until the bubble is at the center of inclinometer leveling instrument. The scale value which the little pointer tip points to is the dip angle.

Expression methods of strata attitude are as follows.

1) The azimuth angle representation method only records inclination and inclination angle, which is suitable for field records, geological reports and profile diagrams. For example, SE125°∠45° (or 125°∠45°), the former is the dip azimuth value, and the latter is the dip angle value, that is, the dip direction is 135° south-east, and the dip angle is 45°.

2) The quadrant angle representation is marked with east, south, west, and north, and the direction of true north or true south is 0°, and the direction of true east or west is 90°. The quadrant angle records the quadrants of the strike line or the trend line, the acute angle between them and the north or south direction. Generally, the strike, dip angle and tilt quadrant are noted. For example: N75°W/35°S, read as 75° north-west

direction, 35° dip angle, and southward dip.

3) Symbolic notation is used in geological maps and horizontal section maps. Commonly used symbols are:

 Long line means strike, short line with arrow means dip direction, while number means dip angle.

—— Horizontal strata (dip angle is 0°~5°)

—— Vertical strata (dip angle is >85°, arrow points to younger strata)

—— Reverse strata (arrow points to reverse
30° direction, the older strata)

3.3.3 Outcrop Characteristics and Contact Relationships of Strata

(1) Outcrops of strata

Outcrops: the parts of strata, ores, rocks, underground water, gas which are above the ground. Natural outcropping is called natural outcrops, while ground exposes caused by various construction projects are called artificial outcrops.

Outcrop lines: The cross lines of strata (including faults and joints) and ground, whose states are decided by the attitudes and ground fluctuation (topography) (Fig. 3-7).

Outcrop width: The distance between the top and bottom of strata on the ground or in the maps (planes), along the vertical direction of strata.

(2) Classification and interpretation of strata in geological maps according to dip angles

1) Horizontal strata (Fig. 3-8a and Fig. 3-9 I): dip angle<5°. The bending shape of the boundary line of this type of outcrop changes with the terrain contour, and bending shape and terrain contour are parallel or coincident. At this time, the thickness of the horizontal rock formation is the difference in elevation between the top and bottom of the rock formation.

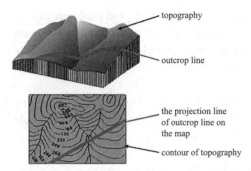

Fig. 3-7 Outcrop lines of strata

2) Vertical strata (Fig. 3-8b and Fig. 3-9 Ⅱ): dip angle>85°. The strata boundary is not affected by topographical change and extends straight along the strike. Only when the rock strata changes or bends, it turns or bends accordingly.

3) Inclined strata (Fig. 3-8c and Fig. 3-9 Ⅲ): dip angle among 5°~85°. The shape of the outcrop line is more complicated, it is a curved extension of the geological boundary line and the contour line with a certain rule that the line is "V"-shaped, and there are three situations:

① When the rock strata tend to be opposite to the slope of the ground, the tip of the "V" shape at the ridge points to the foothills. In the valley, the tip of the "V" shape points to the upper reaches of the valley, but the degree of curvature of the boundary line of the rock strata is smaller than the degree of curvature of the topographic contour line (Fig. 3-10a).

② When the inclination of the rock strata is consistent with the inclination of the ground, and the inclination of the rock strata is greater than the slope angle of the terrain, the bending direction of the rock strata boundary is opposite to the bending direction of the contour line. In the valley, the tip of the "V" shape points to the lower reaches of the valley. On the ridge, the "V" tip points to the uphill slope of the ridge (Fig. 3-10b).

③ When the inclination of the rock strata is consistent with the direction of the ground inclination, and the inclination angle of the strata is smaller than the slope angle of the terrain, the bending direction of the boundary line of the strata is the same as the contour line. In the valley, the "V" tip points to the upper reaches of the valley; On the ridge, the "V" tip points to the downhill slope of the ridge, but the bending tightness of the boundary

(a) Horizontal strata (b) Inclined strata (c) Vertical strata

Fig. 3-8 The strata are classified by their attitudes

Ⅰ—Horizontal strata; Ⅱ—Vertical strata; Ⅲ—Inclined strata

Fig. 3-9 Their outcrops states and expression in the plan

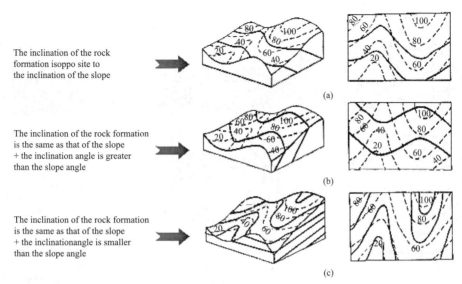

The inclination of the rock formation isoppo site to the inclination of the slope	(a)
The inclination of the rock formation is the same as that of the slope + the inclination angle is greater than the slope angle	(b)
The inclination of the rock formation is the same as that of the slope + the inclinationangle is smaller than the slope angle	(c)

Fig. 3-10 The characteristics of monoclinic rock formation in geological maps

line of the rock strata is greater than that of the terrain contour (Fig. 3-10c).

Table 3-3 summarizes the "V" rule for inclined rock formations.

(3) Contour

Contour refers to close curves connected by adjacent points with equal elevation. Landforms are similar, while topographies are reverse. The higher the mountain head is toward the center; the opposite is true for depressions.

Ridge: The highest ridgeline of a highland extending in one direction (Fig. 3-11, Fig. 3-13).

Valley: A depression between two ridges, the lowest points of which connect (Fig. 3-12, Fig. 3-14).

Steep cliff: The slope is above 70° (Fig. 3-15).

Precipice: Vertical cliffs (Fig. 3-15).

Cliff: The mouth of the cliff is inclined to the outside of the steep wall and hangs in the air (Fig. 3-15).

(4) Contact relationship

The characteristics of the interface and the contact relationship of the rock strata formed in different periods can reflect the development and evolution of various tectonic movements and paleo geographical environments in terms of space and time. The strata contact relationship includes conformity, parallel unconformity contact, angular unconformity contact between

The "V"-shaped rules for inclined rock strata   Table 3-3

Three cases of inclined rock formation	Relationship between rock dip and slope angle	V-shaped rules	Characteristics of outcrops
The dip direction of rock formation is opposite to the slope	—	opposite and same	Curvature is less than contour
The dip direction of rock formation is same to the slope	Inclination > Slope angle	same and opposite	The outcrop line is opposite to the contour line
The dip direction of rock formation is same to the slope	Inclination < Slope angle	same and same	Curvature is more than contour

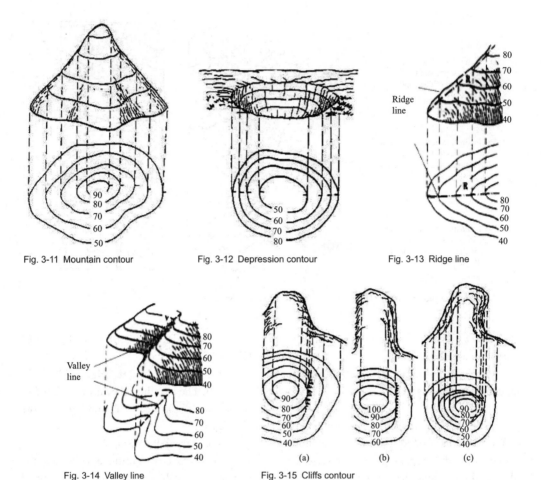

Fig. 3-11 Mountain contour   Fig. 3-12 Depression contour   Fig. 3-13 Ridge line

Fig. 3-14 Valley line   Fig. 3-15 Cliffs contour

sedimentary rocks, and sedimentary contact and intrusive contact between magmatic rock and surrounding rock.

1) Conformity:

The conformity means that strata were formed in a stable sedimentary environment. The dip angle and strike of strata are basically the same.

2) Unconformity (Fig. 3-16):

Unconformity means that sedimentation is interrupted, eroded or deformed. Unconformities range from slight erosion fractures to strong angular inconsistencies in the bedding. Unconformity refers to a period when there is no record of rock accumulation. The time intervals cannot be determined.

① Parallel unconformity (disconformity contact): This type of contact relationship appears roughly parallel stratigraphic boundaries on the geological map. The strata of certain age

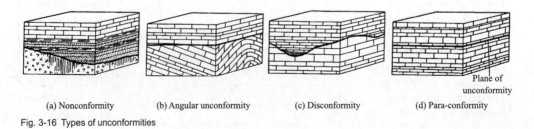

(a) Nonconformity   (b) Angular unconformity   (c) Disconformity   (d) Para-conformity

Fig. 3-16 Types of unconformities

are missing between the two sets. The interface between the two sets of rock strata is called the denuded surface, or the unconformity surface.

② Angular unconformity: Some strata are missing between the adjacent new and old strata, and their attitudes are different from each other. They are angled obliquely, and the strata of some geological age are significantly missing in the middle.

3) Intrusive contact is the contact relationship formed by the intrusion of magma into the previously formed rock layers (Fig. 3-17a). There are often baking alteration or thermal metamorphism, easy to weathered and broken, near the interspersed wall rock contact surface. Later intrusive magma is often mixed with surrounding rocks to form the usually called xenoliths.

4) Sedimentary contact is the contact relationship between sedimentary rocks and igneous rocks (Fig. 3-17b, Fig. 3-18). After the weathering and erosion, of previously formed magmatic rock mass suffered, a new rock layer is formed through deposition. The formation process of sedimentary contact is that when the intrusion body is formed, the crust rises and undergoes denudation, the surrounding rock above the intrusion body is denuded, and then the crust descends and accepts deposition on the denuded surface to form a new stratum.

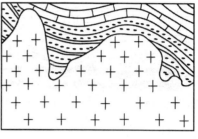

(a) Intrusive contact

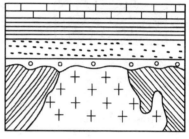

(b) Sedimentary contact

Fig. 3-17 The contact relationship between magmatic rock and sedimentary strata

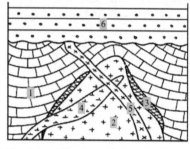

1—limestone, earliest; 2—granite, later than limestone; 3—skarn, same period with granite; 4—diorite, later than granite; 5—diabase, later than diorite; 6—conglomerate, latest

Fig. 3-18 Contact relationships of igneous rocks and sedimentary rocks

3.4 Geological Structure: Folds

3.4.1 Conceptions

Geological structure is the study of the permanent deformation and rock failure created by the changes in stress through geologic time.

Geological structures may be subdivided further into two groups:

(1) Brittle structures: recording the brittle-elastic failure of rocks in the past. Faults and joints fall in this broad category.

(2) Ductile structures: preserving the permanent visco-plastic deformation of rock throughout geological time. Folds and metamorphic foliations are the expression of this type of structure.

Folds under high temperature and pressure are usally plastic deformation of layered rocks at low strain-rates. (Fig. 3-19, Fig. 3-20).

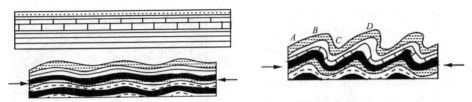

Fig. 3-19 The formation of rock strata under lateral extrusion stress

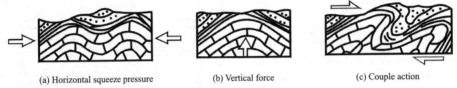

(a) Horizontal squeeze pressure   (b) Vertical force   (c) Couple action

Fig. 3-20 Mechanical causes of the folds

3.4.2 Types of Folds

Anticlines and synclines are symmetrical if both limbs are arranged equally along the axial plane so that the dips on opposing flanks are the same; otherwise, they are asymmetrical (Fig. 3-21).

(1) Anticline (Fig. 3-22)

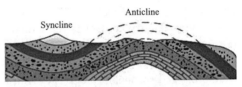

Fig. 3-21 Anticline and Syncline

The deformation of strata bends upward, the strata at core of anticlines are oldest, while strata at two limbs/wings become younger in order from inner (core) to outer (boundary).

(2) Syncline (Fig. 3-23)

The deformation of strata bends downward; the strata at core of anticlines are youngest, while strata at two limbs/wings become older in order from inner (core) to outer (boundary).

3.4.3 Elements of Folds

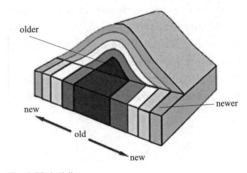

Fig. 3-22 Anticline

Core: the center part of folds. Usually refers to a rock layer that folds out of the most central part of the surface (Fig. 3-24).

Wing/Limb: the side part of folds. Usually, the exposed rock layers on both sides are symmetrical. When the anticline is connected to the syncline, one of the wings is shared.

Axial plane: the plane which divides the fold into two equal parts through core.

Axis: the cross line of axial plane and any other plane.

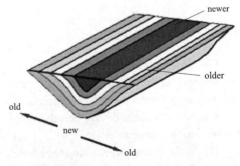

Fig. 3-23 Syncline

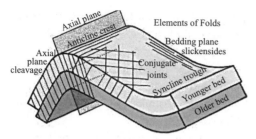

Fig. 3-24 Elements of some anticline

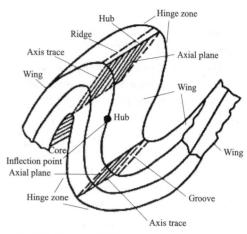

Fig. 3-25 Elements of folds

Pivot: the connection line of maximum bending points in the same layers.

Ridge: the connection line of anticline peak points.

Groove: the connection line of syncline lowest points.

Inflection point (Fig. 3-25): The fold surfaces of adjacent anticlines and syncline common wings are often curved in an "S" or inverted "S" shape.

Wing angle: The maximum angle between the wing rock formation and the horizontal plane is called the wing angle. The size of the wing angle reflects the intensity of folds.

Interlimb angle: the inner angle between two wings on an orthogonal profile. The angle between the wings of the arc-shaped fold is the angle between the tangents through the two inflection points on the two wings. Interlimb angle is as large as 180°, as small as 5° or less.

Hinge zone: the bending part of the fold from one wing to the other wing, the converging part of the two wings. Its shape is often round and smooth, but it can also be pointed, box-shaped, flexed, and so on.

3.4.4 Classification of Folds

(1) According to the orientation of axial (Fig. 3-26 and Fig. 3-27)

Upright: vertical axial plane, symmetric limbs, reverse dip direction, dip angles are similar with no more than 5° difference.

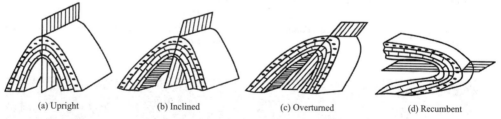

Fig. 3-26 Folds classification according to the attitude of axial plane

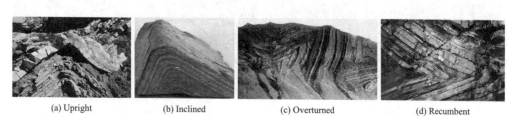

Fig. 3-27 Folds classification according to the attitude of axial plane in the field

Inclined/Asymmetric: inclined axial plane, asymmetric limbs, reverse dip direction, dip angles are unequal.

Overturned: inclined axial plane, same-direction inclined limbs, and one limb overturned.

Recumbent: near horizontal axial plane, one inverted limb.

(2) According to the attitude of fold hinge

1) Horizontal folds: The hub extends almost horizontally and the strata on the two wings are roughly parallel and symmetrically distributed. (Fig. 3-28a, Fig. 3-29)

2) Plunging folds: The hinge is slumped to one end and the strikes of the two wings are arc-shaped and closed. For anticlines, the closed tip points to the tilting direction of the hub. For synclines, the closed tip points to the rising direction of the hub. (Fig. 3-28b, Fig. 3-30)

(3) According to plane forms

1) Linear folds: the ratio of length to width is >10:1.

2) Short axis folds: the ratio of length to width is 3:1~10:1.

3) Dome and Basin: The extension is very short; the length of the fold is less than 3 times of the width. If the fold is an anticline, it called a dome. If the fold is a syncline, it called a tectonic basin. (Fig. 3-31)

(4) According to thickness of folding layers

1) Parallel folds

Parallel or concentric folds (Fig. 3-32a)

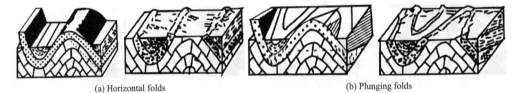

(a) Horizontal folds   (b) Plunging folds

Fig. 3-28 Horizontal folds and plunging folds

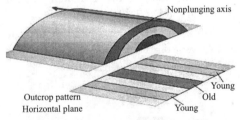 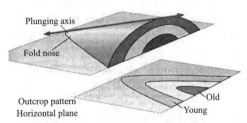

Fig. 3-29 Horizontal folds (Geological boundaries are parallel lines)

Fig. 3-30 Plunging folds (Geological boundaries are closed curves)

(a) Dome   (b) Basin

Fig. 3-31 Dome and basin (Eye of the Sahara)

are those where the strata have been bent into more or less parallel curves in which the thickness of the individual bed remains the same. It can be observed, because the thickness of the beds remains the same on folding, the shape of the fold changes with depth and, gradually fade out. Parallel folding occurs in competent (relatively strong) beds that may be interbedded with incompetent (relatively weak, plastic) strata.

2) Similar folds

Similar folds (Fig. 3-32b) are those that retain their shape with depth. The curvature of each surface is the same, but there is no common center of curvature. Similar folds are developed in incompetent strata. However, exactly similar folds are rare in the nature, for most change their shape to some degree along the axial plane. Most folds exhibit both the characteristics of parallel and similar folding. Fig. 3-33 shows some types of folds.

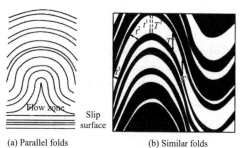

(a) Parallel folds   (b) Similar folds

Fig. 3-32 Parallel folds and similar folds

3.4.5 Field Identification Methods of Folds

In general, people tend to think that an anticline will become mountain and syncline will become valley. However, the actual situations are much more complicated than this. Due to the formation of the folds, the multi-layer shafts are extremely susceptible to fracture and damage caused by structural

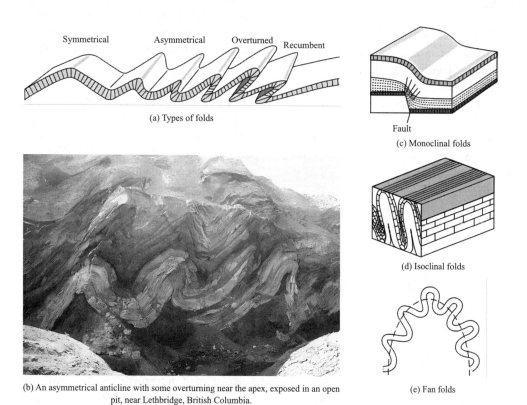

(a) Types of folds

(b) An asymmetrical anticline with some overturning near the apex, exposed in an open pit, near Lethbridge, British Columbia.

(c) Monoclinal folds

(d) Isoclinal folds

(e) Fan folds

Fig. 3-33 Some types of folds

extrusion. Suffering from long-term denudation under a certain external force, it will be gradually razed to the ground, and even developed into a valley. Syncline mountains and anticline valleys are more common in the field (Fig. 3-34). Therefore, the undulation of the terrain cannot be used as the main indicator to identify fold structure.

(1) Traverse method

The traverse method refers to the investigation and observation along the vertical direction of strata strike of selected roads. It is helpful to grasp the attitude, sequence, old and new relation of strata.

(2) Tracing method

The tracing method refers to the investigation and observation along the parallel direction of strata strike. It is helpful to grasp the extending and variety of strata.

3.4.6 Engineering Significance of Folds

Folds, widespread in the nature, are very important for mining, underground water locating and engineering construction.

For example:

(1) Just because the symmetric characteristics of limbs, if sedimentary mine strata are found on one side of fold, then the same mine strata also exist on the other side. In addition, petroleum is often stored at the top of the anticline.

(2) Due to the horizontal compression of the rock in the fold core, many cracks are

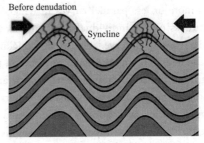
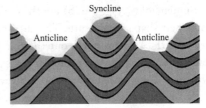

Fig. 3-34 Before and after denudation

generated, which can directly decrease the integrity and strength of rock mass, and facilitate the development of karst in limestone areas. If projects i.e., roads, bridges, dam sites, and tunnels are arranged in the core, more attention must be paid to preventing rock collapse, water leakage and water gushing.

(3) Since the rock mass at core of anticline tends to fracture frequently, so the reservoirs and undergrounding projects should avoid being applied at the cores of folds, while limbs are reasonable site choices.

(4) Tunnels and other underground projects should generally be arranged at the folds of the wing, because uniform rock formations are conducive to realizing tunnel stability.

3.5 Geological Structure: Fractures

The rock layers that make up the Earth's crust can be deformed under the action of tectonic stress. When the deformation reaches a certain level, the continuity and integrity of the rock formation will be destroyed, resulting in various forms of fractures, which are called fracture structures.

According to displacement degree along the fracture surface of the rock blocks on both sides of the fracture, the fracture structure is divided into two categories: joints and faults.

3.5.1 Joints

Joints refers to small fracture structures where there is no significant relative displacement of the rock blocks on both sides after the rock mass is fractured under force. The fractures surface of the joint is called the joint plane. Joints are widely distributed in almost all rock formations. Joints often divide rock formations into rock blocks of different shapes and sizes.

(1) Types of joints

1) According to the origin of joints

① Origin joints: formed during rock formation. (Fig. 3-35)

② Structural joints: caused by tectonic movements. (Fig. 3-36)

③ Secondary joints: caused by unloading, weathering and blasting, etc. (Fig. 3-37)

2) According to the relation between joints and attitude of strata. (Fig. 3-38)

① Strike joints: The joints are roughly parallel to the strike of the strata.

② Dip joints: The joints are roughly perpendicular to the strike of the strata.

③ Oblique joints: The joints obliquely intersect with the strike of the strata.

④ Bedding joints: The joint plane is roughly parallel to the strata.

3) According to the mechanical properties of joints formation.

① Shear joints (Fig. 3-39):

The fracture surface formed by the shear (torsion) stress of the rock is called the shear joints, which are often associated with folds and faults.

② Tension joints (Fig. 3-40):

The fracture surface formed by the tension stress of the rock formation is called tension joints. When the rock formation is compressed, the shear joints that occurred first on the rock layer will develop into zigzag tension joints. In fold rock formations, tension

Fig. 3-35 Origin joints (columnar joints of basalt)

Fig. 3-36 Structural joints

Fig. 3-37 Secondary joints

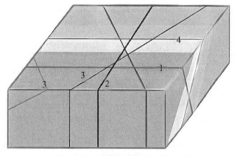

1—strike joint; 2—dip joint; 3—oblique joint;
4—strike of strata

Fig. 3-38 The relationship between the joints plane and the attitude of strata

Fig. 3-39 Shear joints

Fig. 3-40 Tension joints

joints consistent with the direction of the fold axis are often formed at the top of the bend.

Comparison of shear joints and tension joints is shown in Table 3-4.

4) According to the relation between joints and fold axis.

Longitudinal joint: The joint direction is roughly parallel to the fold axis. (Fig. 3-41 part A)

Oblique joint: The joint direction obliquely intersects with the fold axis. (Fig. 3-41 part B)

Transverse joint: The joint direction is almost perpendicular to the fold axis. (Fig. 3-41 part C)

5) According to the opening degree of joints (Fig. 3-42)

Wide joints: width >5 mm
Open joints: width 3~5 mm
Slight open joints: width 1~3 mm
Closed joints: width <1 mm

(2) Survey and statistics of joint

1) Selection of observation points

First, to determine the scope and level of detail of the investigation according to the investigation nature and tasks. The number and density of observation points depend on the specific situation and task.

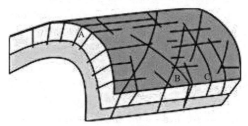

A—longitudinal joints; B—oblique joints; C—transverse joints
Fig. 3-41 Development characteristics of the joints

2) Survey and its main items

In order to reflect the distribution of joints and their influence on the stability of rock mass, it is necessary to conduct field surveys, and use statistical graphs and other methods to organize and analyze survey data.

Select representative outcrops of bedrock during the investigation, and investigate joints in a certain area. The selection of observation points should be based on the situation and

Comparison of shear joints and tension joints　　　　Table 3-4

Characteristics	Shear joints	Tension joints
Attitude	Stable, long	Unstable, short
Joint plane	Straight and smooth, with slickensides	Coarse and uneven, no slickensides
Relation with graves	Cutting through gravels	Bypassing gravels
Combined forms	"X" shape	Irregular tree branches
Joint plane composition	Feather micro fracture surfaces	Sometimes with lateral fissures

Fig. 3-42 Joint development with different degrees of opening

tasks of the survey to determine the scope and level of the survey. Its main items are as follows:

A—joints origin and mechanical properties;
B—joints set numbers, density and their attitudes;
C—joints open degree, extending distance and coarse degree of joints plane;
D—joints infilling substance, thickness and water content;
E—classification of joints development.

The degree of joints development can be used to quantitatively reflect the degree of joints development in the rock with joint density and joint rate.

① Joint density refers to the numbers of joints per unit length of the vertical joint direction.

Formula:

$$u = \frac{s}{l} \qquad (3\text{-}2)$$

In the formula: s is the number of joints; l is the measured length.

② Joint rate is the percentage of the area occupied by cracks per unit area.

Formula:

$$K_{tp} = \frac{\sum I_b}{A} \qquad (3\text{-}3)$$

In the formula, $\sum I_b$ is the total area of the fissures; A is the measured rock outcrop area.

The joint rate indicates the degree of development of joints in the rock. The larger the joint rate, the more developed the joints in the rock, the greater the compressibility and water permeability of the rock, and the lower the shear strength (Table 3-5).

3) Collation of joint observation data

The collation, statistics and structural analysis of joint data generally adopt the form of graphs, mainly including joint rose diagrams, joint stereographic projection pole

Classification of joint development   Table 3-5

Development degree	Basic characteristics
Not developed	1~2 groups, regular, structural joints, spacing > 1 m, mostly closed joints, rock mass is cut into big blocks
Relative developed	2~3 groups, relative regular, "X" shape, structural joints, spacing > 0.4 m, mostly closed joints, some open joints, rock mass is cut into big blocks
developed	>3 groups, irregular, "X" shape, mostly structural or weathering joints, spacing < 0.4 m, mostly open joints, part with infillings, rock mass is cut into blocks
Well developed	>3 groups, disordered, mostly weathering or structural joints, spacing > 0.4 m; open joints, some wide joints, with infillings, rock mass is cut into little blocks

diagrams, joint stereographic projection contour maps, and conjugate joint stereographic projection to solve principal stress azimuth maps (Table 3-6).

The joint rose diagram is easy to compile, and used to reflect the trend of joint attitude. Among them, the rose diagram of joint strike mainly reflects the arrangement and changing trend of joint strike (Fig. 3-43). The rose diagram of joint dip mainly reflects the arrangement and change trend of the joint dip. The rose diagram of the joint dip angle mainly reflects the changing trend of the joint dip angle (Fig. 3-44).

4) Influence of joints on projects

① The joints can destroy the integrity of the rock, creates favorable conditions for weathering, and promotes the weathering speed of the rock.

② The joints can reduce the strength of the rock, the bearing capacity and the stability of the foundation.

③ The existence of joints is conducive to excavation and quarrying, but can affect the effect of blasting operations.

④ The joints can facilitate groundwater passage, which enhances the permeability of the rock mass and accelerates the dissolution of soluble rock, which will cause water gushing during construction (Fig. 3-45).

⑤ Rock formations with well-developed joints are good places of water supply sources.

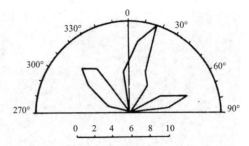

Fig. 3-43 The rose illustration of joints strike

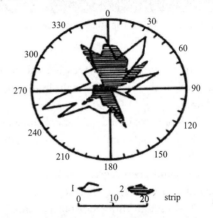

Fig. 3-44 The rose illustration of joints dip direction and dip angle

Fig. 3-45 Dangerous rocks formed by joints sets

The form of joint observation and record   Table 3-6

Point number and location	Stratigraphic age, horizon and lithology	Attitude and structural location of rock formations	Joint attitude

Joint system and its mechanical properties	Joint staging and supporting	Joint density	Joint surface characteristics and fillings	Remarks

3.5.2 Faults

Faults are planar discontinuity surfaces with significant displacement (Fig. 3-46).

(1) Fault elements (Fig. 3-47)

1) Fault plane: The fractures plane where the rock blocks on both sides are displaced relative to each other, and it be a plane or a curved surface. The dip angle and strike of a fault plane can be described in the same way as those of a bedding plane. Sometimes the movement on both sides of fault does not occur along one surface, but along a fractures zone composed of many fractures' surfaces. This zone is called a fault fracture zone or a fault zone.

2) Fault wall: A fault wall is a rock block or a rock mass that moves relatively on both sides of a fault plane. (Fig. 3-48)

① Footwall

The footwall is located below the fault plane.

② Hanging wall

The hanging wall is the above the fault plane.

Where the displacement along a fault is vertical, it should be described in terms of direction, such as the east or west side of the fault. Divided by the direction of relative movement, the terms downthrow and upthrow refer to the relative movement of strata on opposite sides of the fault plane.

3) Fault line: The line of intersection between the fault plane and the ground, which reflects the direction in which the fault extends on the surface. It can be a straight line or a curve one.

4) Fault distance: The displacement distance of the same point in a rock formation after being broken by a fault (see "ab" in Fig. 3-49).

In general (unless the fault is vertical), there are vertical and horizontal component of movement. The vertical movement is called the throw (see "a-m" in Fig. 3-49) and the horizontal component the heave (see "b-m" in Fig. 3-49).

Fig. 3-46 Faults in the field

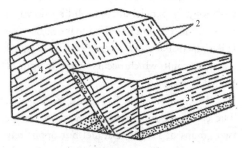
1—fault plane; 2—fault lines; 3—footwall; 4—hanging wall
Fig. 3-47 Fault elements

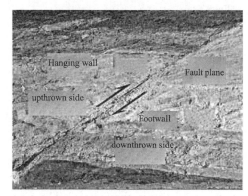
Fig. 3-48 Fault wall

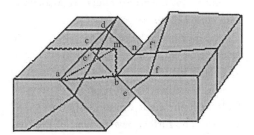
Total displacement (a-b); heave (a-m); throw (b-m)
Fig. 3-49 Fault distance

(2) Classification of faults

1) According to relative movement direction of two walls:

① Normal Fault (as shown in Fig. 3-50)

In normal faulting, the hanging wall moves down relative to the footwall. The fault plane usually makes a high angle with the ground (>45 degrees). Normal faults are associated with crustal tension.

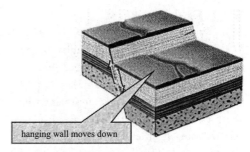

Fig. 3-50 Normal Fault

(a) Stepped fault (Fig. 3-51): The combination of faults formed by many normal faults with roughly parallel attitude sliding down in the same direction in a stepped manner. Because it is shaped like a descending step, it is called a stepped fault, which occurs at the margin of faulted basins.

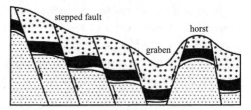

Fig. 3-51 Stepped faults, graben and horst

(b) Graben (Fig. 3-51 and Fig. 3-52): The combination of faults formed by two or two groups of roughly parallel but oppositely inclined faults. The two sides of the rock mass between the fault planes fall relatively, and the rock masses on both sides rise along the fault plane.

(c) Horst (Fig. 3-51 and Fig. 3-52): The combination of faults formed by two or two groups of faults with roughly parallel strikes but opposite inclination, and such type is opposite to the aforementioned graben. The rock masses on both sides of the fault plane descend along the fault plane, while the rock mass in the middle rises along the fault plane.

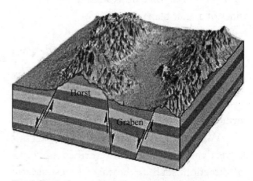

Fig. 3-52 Graben and horst

② Reverse Fault (Fig. 3-53)

In reverse faulting, the hanging wall moves up relative to the footwall. The fault plane usually makes a low angle with the ground (<45°). Reverse faults are associated with crustal compression.

(a) Thrust fault: refers to a high-angle reverse fault with a fault dip angle greater than 45°(Fig. 3-54).

(b) Overthrust fault: refers to the reverse fault with the fault dip angle between 25° and 45°. The further development of inverted folds often forms such faults. The scale of the overthrust fault is generally large, and its strike is roughly parallel to the fold axis.

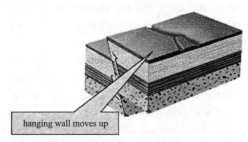

Fig. 3-53 Reverse Fault

(c) Crushing fault: refers to the reverse fault with a dip angle of less than 25°.

(d) Shingled structure: refers to the combination of a series of reverse faults with roughly the same attitude and arranged in parallel. The hanging wall rock blocks of each fault are uplifted, and they are successively

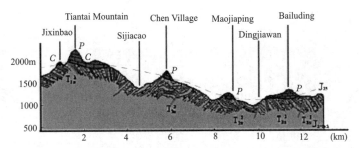

Fig. 3-54 Thrust nappe structure in Pengxian County, Sichuan Province, China

overlapped in a tile-like shape on the section (Fig. 3-55).

③ Strike-slip Fault (Fig. 3-56)

In strike-slip faulting, the two walls move either to the left or to the right relative to the other. Strike-slip faults are associated with crustal shear and are also known as wrench faults.

2) According to relation between fault attitude and strata attitude

① Strike fault: The strike of the fault is basically the same as the strike of the rock formation (F_1 in Fig. 3-57).

② Dip fault: The strike of the fault is basically orthogonal to the strike of the rock formation (F_2 in Fig. 3-57).

③ Oblique fault: The strike of the fault obliquely intersects the strike of the rock strata (F_3 in Fig. 3-57).

④ Bedding faults: The attitude of fault plane is basically the same as the primary geological interface such as strata bedding.

3) According to relations between fault strike and fold strike

① Longitudinal fault: the fault direction is parallel to the fold axis (F_1 in Fig. 3-58).

② Traverse fault: the fault direction is perpendicular to the fold axis (F_2 in Fig. 3-58).

③ Oblique fault: the fault strike obliquely intersects the fold axis (F_3 in Fig. 3-58).

4) According to mechanical properties of fault

① Compression fault: Reverse faulting (Fig. 3-59)

The minimum principal stress is vertical. The maximum and intermediate principal stresses are horizontal. The shear fracture

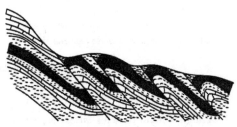

Fig. 3-55 Shingled structure

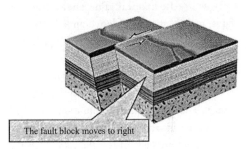

Fig. 3-56 Strike-slip Fault

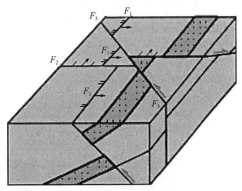

F_1—strike fault; F_2—dip fault; F_3—oblique fault
Fig. 3-57 Strike fault, dip fault and oblique fault

(fault) makes an angle of less than 45° with the major (maximum compressive) principal stress direction.

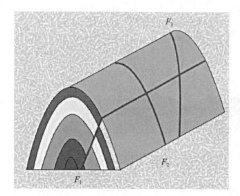

F_1—longitudinal fault; F_2—traverse fault; F_3—oblique fault

Fig. 3-58 Longitudinal fault, traverse fault and oblique fault

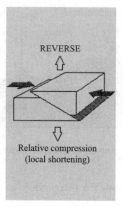

Fig. 3-59 Relative compression

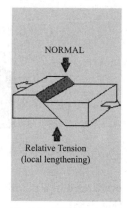

Fig. 3-60 Relative tension

② Tension fault: Normal faulting (Fig. 3-60)

The maximum principal stress is vertical. The minimum and intermediate principal stresses are horizontal. The shear fracture (fault) makes an angle of less than 45° with the major (maximum compressive) principal stress direction.

③ Shear fault: strike-slip faulting

The intermediate principal stress is vertical. The maximum and minimum principal stresses are horizontal. The shear fracture (fault) makes an angle of less than 45° with the major (maximum compressive) principal stress direction, in this case again it is horizontal. There is no vertical movement. Mostly appear in pairs, with "X" cross shapes.

(3) Fault identification marks

Most of the faults in the natural world are not exposed on the surface due to the erosion damage and coverage. Therefore, it is necessary to identify the existence and type of faults according to strata structure, attendant phenomenon, landforms and hydrogeology, etc.

1) Structural signs (structural lines and geological bodies are discontinuous). The sudden interruption and staggering of geological bodies such as rock formations, ore-bearing layers, rock masses, and fold axes, or geological boundaries, etc. (Fig. 3-61, Fig. 3-62), which indicate that there may be faults. However, it should be distinguished from discontinuities caused by

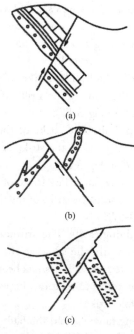

Fig. 3-61 Discontinuous marks caused by the fault

Fig. 3-62 The rock vein is cut off, traction is generated

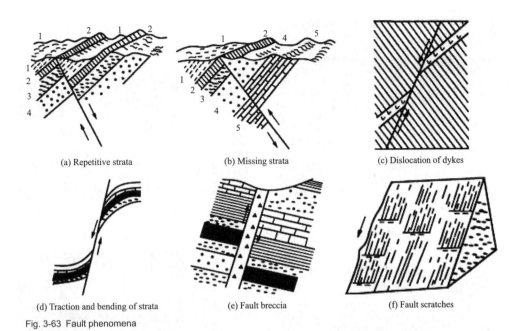

(a) Repetitive strata (b) Missing strata (c) Dislocation of dykes

(d) Traction and bending of strata (e) Fault breccia (f) Fault scratches

Fig. 3-63 Fault phenomena

unconformity interfaces and rock mass intrusive contact interfaces.

2) Missing or duplication of strata. In sloping rock formations, the occurrence of duplication or lack of formations is an important sign to identify the existence of faults. In a certain area, according to the normal stratigraphic sequence, if there are some asymmetric repetitions (Fig. 3-63a), missing (Fig. 3-63b) or dislocation of dykes (Fig. 3-63c), as well as the sudden absence or thickening or thinning of certain stratum, there may be faults.

3) The accompanying structures of the fault refer to the various evidences left on and near the fault planes due to the mutual sliding and friction of the rock blocks on both sides of the faults. Common structures include traction bending, scratches (Fig. 3-64 and Fig. 3-65), steps, friction mirrors, fault breccia (Fig. 3-66),

Fig. 3-64 Scratches on the thrust fault plane

Fig. 3-65 Scratches

Fig. 3-66 Fault breccia

Fig. 3-67 Fault gouge

Fig. 3-68 Channeling bead-like spring water formed by fault activity

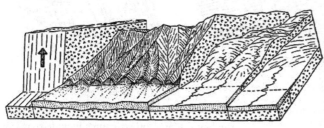

Fig. 3-69 Fault triangle

and fault gouge (Fig. 3-67). These are reliable signs for identifying faults.

4) Geomorphology and hydrological signs. The Fig. 3-68 shows the channeling bead-like spring water formed by fault activity and Fig. 3-69 shows the fault triangle.

(4) The formation age of the fault

The formation age of the fault can be determined according to the cutting relationships between the fault and the stratum. If the fault cuts through a set of strata, the formation age of the fault should be later than the latest stratigraphic age in this set of strata. When the fault is covered by another set of strata, the formation age of the fault is earlier than the oldest strata age in the overlying strata.

(5) Engineering geological evaluation of faults

Generally speaking, faults destroy the integrity of the rock mass, and the shear strength of the fault plane or fracture zone is much lower than that of other parts of the rock mass. As a result, faults can affect engineering construction and project operation from the following aspects.

1) Faults reduce the strength and stability of rock mass. The mechanical strength of the fault fracture zone is low, the compressibility is larger, and large subsidence will occur, which can easily cause the cracks or tilt of the building.

2) In the faulted structural zone, rock mass tends to be broken or failure, and the lithology of the upper and lower walls of the fault may be different. If construction is carried out in the zone, uneven settlement may occur.

3) Collapse or even roof fall can occur easily when tunnel projects are arranged to pass through the fractured zone.

4) Deep weathered troughs and karst development zones are easy to form in sections along the fractured zone. Fault steep slopes or cliffs are mostly in an unstable state and prone to collapse.

5) The fracture zone of the fault structure

is often a good passage for groundwater, and the outburst of groundwater is often controlled by the fault structure. During construction, water gushing will occur when a fault zone is encountered.

6) Under the influence of new crustal movement, new movement may occur in structural fault zone. Because the structural fault zone is a weak zone on the surface of the Earth's crust, if a new crustal movement occurs, there will be new movements in the nearby fault zone, thereby affecting the stability of the buildings.

3.6 Geological Map

3.6.1 Concepts and Types of Geological Map

(1) Geological map concept

A geological map refers to a drawing using prescribed symbols, colors, and various patterns and lines to show geological phenomena of a region, such as stratum, geological structure in a certain reduced proportion. It is an important geological data that needs to be collected and studied in engineering practice.

(2) Types of geological maps

1) General geological map: A basic map showing the strata and lithology and geological structure conditions of a certain area.

2) Structural geological map: The geological map that uses special symbols to indicate the attitude of rock formations, fold shafts and faults, and the attitude and distribution of joints.

3) Quaternary geological map: A geological map showing the lithology, distribution, genesis and age of the Quaternary strata.

4) Geological map of bedrock: A map only reflects the age, lithology and distribution of bedrock before the Quaternary by omitting loose Quaternary sediments.

5) Hydrogeological maps: A map reflects regional hydrogeological data. It can be divided into water content map in rock formation, groundwater chemical composition map, phreatic water level map, comprehensive hydrogeological map and other types.

6) Engineering geological map: Such map is compiled for engineering purposes. It is formed by integrating various engineering geological investigation methods: surveying and mapping, exploration, testing, etc., and the results analysis and comprehensive compiling.

3.6.2 Representation Methods of Geological Conditions on the Geological Map

(1) Geological structures. Geological structures are generally identified based on the legend symbols of the chart (as listed in Table 3-7). If there are no legend symbols, folds can be determined according to the symmetrical distribution relationship between the new and old rock formations. Faults can be identified according to phenomena such as duplication, lack, interruption, width change, or dislocation of rock formations.

(2) Contact relationship of strata

Parallel unconformity. First, check whether the stratigraphic sequence is missing from the legends and the geological map. If it is missing, the upper and lower stratigraphic boundaries are basically parallel (the two sets of attitudes are basically the same), it is a parallel unconformity contact. If there are no missing strata, and the attitudes of the upper and lower strata are the same, it is a conformable contact.

Angular unconformity. If some strata are missing, and the attitudes of the upper and lower strata are not parallel, but intersect at an angle, it is an angular unconformity.

Representation of geological structures in geological map   Table 3-7

Geological structures	Strata characteristics	Display methods	Remarks of illustration
Strata attitude	Horizontal strata		The long line indicates the strike and the short line indicates the dip
	Inclined strata	30°	The long line indicates the strike, the short line indicates the dip, and the number indicates the dip angle of strata
	Vertical strata		The arrow indicates the new rock formation
	Inverted strata		The arrow indicates the tendency after inversion
Folds	Syncline		
	Anticline		
	Reverse tilt		
	Inverted anticline		
Fault	Normal fault		The long line indicates the location of the fault and the direction of the fault plane, the vertical long line with an arrow and the short line indicates the dip direction of the fault, and the number indicates the dip angle, while the short line without an arrow indicates that the disc is a descending one
	Reverse fault		
	Strike-slip fault		The arrows parallel to the long line indicate the relative displacement direction of the two walls of the fault

3.6.3 Geological Section Map and Comprehensive Geological Histogram

(1) Geological section map and its classification

Geological section: Along a certain direction, show the actual (or inferred) section of the geological structure within a certain depth.

Geological section line: The intersection line of a geological section with the Earth's surface or a certain plane.

Geological section map: A map, which shows the geological section. In essence, it is a map showing the geological phenomena and their mutual relations on the geological section under a certain scale.

(2) The method of using geological plan map to cut and draw geological section map

The stratigraphic symbol and chromatogram should be consistent with the geological map.

Steps of using a geological plan to cut and draw a geological section:

1) Determine the orientation of the section line. It is generally required to be perpendicular to the stratigraphic strike line or the geological structure line.

2) Determine the scale. Choose an appropriate scale according to the length of the actual section, so that the drawn section is not too long or too short, and at the same time it can indicate the geological elements of each place. It should be noted that the horizontal scale should be the same as that of the plan. The vertical (elevation) scale can be appropriately larger than that of the plan.

3) Draw the terrain contour (topographic line) according to the selected section orientation and scale. Based on the elevation and horizontal distance, project the intersection of the contour line and the section line on the topographic map on the graph paper, and then connect the adjacent points according to the actual topographic situation, namely the topographic line, and mark the section orientation.

4) Measure each geological content according to the required units and attitude with a protractor, and cast them below the corresponding point of the topographical line (the intersection of the geological boundary and the topographical line).

5) Indicate geological elements with general patterns and codes. Mark the name of the map, legend, scale, section orientation and names of features on the section, etc.

(3) Comprehensive geological histogram

The comprehensive geological histogram, based on the geological investigation data (mainly based on the geological plan and borehole histogram data), adopts the form of a histogram to show the exposed lithology, maximum thickness, contact relationship, etc. of an area from old to new, from bottom to top, in the original order of formation.

Generally, there are stratigraphic age, symbols, lithological patterns, stratigraphic contact types, stratigraphic thickness, and lithological descriptions, but the map does not indicate folds and fractures (joints and faults). The stratigraphic histogram is generally attached to the left of the geological map or serve as a separate map, and its scale can be determined according to the requirements for reflecting the detailed level of the strata and the total thickness of the strata (Fig. 3-70).

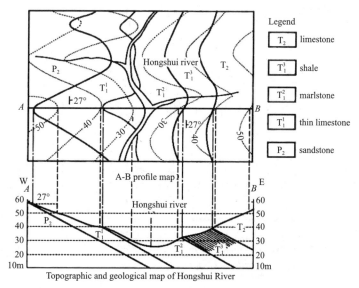

Fig. 3-70 Use the plan to cut the profile diagram

3.6.4 Reading and Analysis of Geological Map

(1) Steps and methods for reading geological map

There are many elements on the geological map, and the lines and symbols are complicated. The general basic principles when reading the map are as follows: from the outside of the map to the map itself.

1) Outside the frame

Map name, scale, legend, map sheet number and latitude and longitude, section map.

2) In the frame: read topography and geomorphology first, then geological items.

① Topography

The general distribution characteristics of the topography in the map and the distribution characteristics of the water system.

② Geological items (first, stratigraphy, second, structure)

The distribution characteristics of the strata in each era and the relationship with the topography.

The distribution characteristics and attitude of horizontal, inclined and upright rock formations.

The contact relationship among strata.

Geological phenomena such as folds, faults, attitudes of magmatic rocks, primary and secondary structures, and structural features of metamorphic rocks.

(2) Example of reading geological map

Takes the following geological map of the Sun Mountain area (Fig. 3-71) as an example to analyze the geological conditions of the area and the analysis is as follows:

1) Scale

The scale of the geological map is 1:15000, i.e., 1 cm on the map represents a field distance of 150 m.

2) Topography

The Sun Mountain (north-south direction) is the highest in the area, and its two sides (east and west) gradually become lower. The Sun Mountain is more than 1100 meters, the ridge is north-south, and the northern peak is more than 1000 meters. There are three river valleys in the area. The largest river valley, flowing from northeast to southwest, is in the southwest with an elevation of about

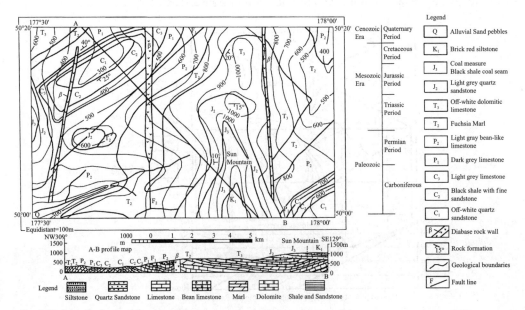

Fig. 3-71 Geological map of Sun Mountain area

300 m, and a wide distribution of Quaternary alluvial deposits on both sides of the valley. The other two lower river valleys are located in the northwest and northeast.

3) Strata and lithology

The strata exposed in this area (from old to new): Paleozoic-Lower Carboniferous (C_1) sandstone, Middle Carboniferous (C_2) shale-sandstone interbedded, Upper Carboniferous (C_3) limestone, Lower Permian (P_1) Pure limestone, Upper Permian (P_2) limestone and shale interbedded; Mesozoic-Middle Triassic (T_2) marl, Upper Triassic (T_3) dolomitic limestone, Middle Jurassic (J_2) Lime sandstone, Upper Jurassic (J_3) shale and sandstone interbedded; Cenozoic-pebbles and sand.

4) Geological structures

The basement fold structure in this area is composed of three folds, all in the NE-SW direction. Among them, the core of the short-axis anticline in the northwest and the southeast (the northwest wing of the anticline is exposed in the map) is composed of gray-white quartz sandstone of the Lower Carboniferous (C_1) strata, and two symmetrically-distributed wings are composed of Middle Carboniferous (C_2), Upper Carboniferous (C_3), Lower Permian (P_1), Upper Permian (P_2), Middle Triassic (T_2), Triassic Upper system (T_3) strata. The open area between the two short-axis anticlines is a syncline with the gray-white dolomite limestone of the Upper Triassic (T_3) as the core. The two wings are symmetrically distributed in the Middle Triassic (T_2) and Upper Permian. Strata (P_2), Lower Permian (P_1), Upper Carboniferous (C_3), Middle Carboniferous (C_2), Lower Carboniferous (C_1); the two wings have a gentle dip angle of about 20°.

There are two groups of faults in the area. One group consists of three north-south extensional faults, all of which are intruded by diabase magma and have formed diabase walls or dikes, and the middle fracture remains a section (F_2) that has not been intruded by diabase; the other group are NE-SW strike, such as F_1, the fault is in line with the basement fold axis in the area, and its dip angle is almost upright, and there is no obvious displacement of the rock formations on both sides of the fracture surfaces.

5) Contact relationships

There are no missing strata between the Carboniferous and Permian strata, and their rock formations are consistent in attitude and are in conformable contact.

The attitudes of strata between the Permian and Triassic strata are same, but the Lower Triassic strata are missing, so the two are in parallel unconformity contact. The Jurassic in the map is in contact with the older strata of Carboniferous, Permian, and Upper Triassic, and the strata are obliquely intersecting, which is an angular unconformity contact. The Quaternary and the old strata are in angular unconformity contact.

Diabase intrudes into Carboniferous, Permian and Middle and Upper Triassic strata along three north-south tensile faults, as a result, the three diabase rock walls or dykes exposed in the area are in intrusive contact with the Carboniferous, Permian, and Middle-Upper Triassic strata, while they are in sedimentary contact with the Middle-Upper Jurassic and Lower Cretaceous strata. Therefore, the geological age of the formation of diabase should be after the Triassic and before the Middle Jurassic. The Lower Jurassic (J_1) strata are missing in the area, and the upper Triassic (T_3) and Middle Jurassic (J_2) strata are in angular unconformity contact. Therefore, during the Early Jurassic (J_1) period, a large-scale tectonic movement occurred in this area—the Indosinian movement which resulted in the basic fold structure and north-south tensile fault in this area.

Quiz

1. What is geological process? What are the types of geological process?

2. How to determine the relative geological time?

3. Briefly explain the definition of the three elements of rock strata attitude and their representation methods.

4. How to use the relationship between rock strata dip and slope direction to infer the outcropping position of the rock strata?

5. Briefly describe the concept, types, characteristics and identification marks of folds.

6. Briefly describe the classification of fractures by their origins and their mechanical properties.

7. Briefly describe the concepts, elements, common types and characteristics of faults.

8. How to draw the rose illustration of joints strike and joint dip direction? How to use the rose illustration to identify the dominant fractures in the survey area?

9. Discuss the engineering significance of folds, faults and joints.

10. How to read and analyze the geological map?

Chapter 4
Engineering Geological Characteristics of Rock and Rock Mass

4.1 Rock and Rock Mass

Rock is a good natural foundation for buildings, and conventional building material. The engineering properties of rock and rock mass have a stake in the safety and stability of buildings. Their engineering properties include physical, hydraulic, and mechanical ones, etc., closely related with the geological characteristics, water conditions, weathering and other factors.

Rock mass refers to a discontinuous, heterogeneous and anisotropic geological body composed of various types of rocks containing structural planes within a certain engineering scope. Rock mass is formed in a long geological history, and occurs in a certain geological environment (i.e., geo-stress, groundwater, ground temperature). The engineering properties of rock masses depend first on the properties of various structural planes, and then the rock properties that make up the rock mass.

4.2 Physical and Hydraulic Properties of Rock

4.2.1 Physical Properties of Rock

(1) Density and Gravity of Rock

The density refers to the mass per unit volume of the rock (g/cm^3 or kg/m^3). The density of rock containing no water voids is called the dry density. The density with voids filled with water is called the saturated density of the rock.

In engineering construction practice, specific gravity is usually used to indicate rock density, that is, the ratio of the weight of the solid part of the rock per unit volume to the weight of the same volume of water (4℃).

The specific gravity could be defined as

$$\gamma = \frac{W}{V} \qquad (4\text{-}1)$$

Where W represents the weight of the rock specimen (kN); V the volume of the rock specimen (including pore volume) (m^3).

Depending on the water content of the rock, the gravity can be divided into natural gravity (γ), dry gravity (γ_d) and saturated gravity (γ_w).

(2) Porosity

Rock porosity refers to the ratio of the volume of rock pores and fractures to the total volume of the rock, namely

$$n = \frac{V_n}{V} = \frac{G - \gamma_d}{G} \times 100\% \qquad (4\text{-}2)$$

Where V is the volume of rock (m^3); V_n the total volume of rock pores (m^3), and γ_d the dry weight (kN/m^3).

(3) Void Ratio

The void ratio (e) of rock refers to the void volume V_v to the solid volume V_s. According to the interrelationship of the three-phase state in the rock, the void ratio e and porosity n have the following relationship.

$$e = \frac{n}{1-n} \qquad (4\text{-}3)$$

4.2.2 Hydraulic Properties of Rock

The hydraulic properties of rocks refer to the properties exhibited when rocks interact with water, which usually include rock water absorption, water permeability, softening and freezing resistance.

(1) Water Absorption

The water absorption capacity of rocks under certain test conditions is called rock water absorption. It depends on the openness and closing of volume, and distribution of rock pores. Rock water absorption is usually characterized by water absorption, saturated water absorption and saturation coefficient.

Rock water absorption (w_a) refers to the ratio of the mass (m_o) of the rock specimen under an atmospheric pressure and room temperature for 48 hours to the mass (m_s) of the dry rock specimen. Formula:

$$w_a = \frac{m_o - m_s}{m_s} \times 100\% \quad (4\text{-}4)$$

The saturated water absorption rate (w_{sat}) of the rock refers to the ratio of the mass of water (m_p) absorbed by the rock specimen under high pressure (15 MPa) or vacuum conditions to the dry mass of the rock specimen (m_s), expressed as a percentage. Reflecting the degree of pore development, it can be used to indirectly determine the frost resistance and weathering resistance of the rock.

Saturated coefficient of rock (k_s) refers to the ratio of rock water absorption (w_a) to the saturated water absorption rate (w_{sat}). Formula:

$$k_s = \frac{w_a}{w_{sat}} \quad (4\text{-}5)$$

(2) Water Permeability

The ability of a rock to be penetrated by water is called rock permeability. Water can penetrate only along connected pores. The water permeability of rock can be indicated by the permeability coefficient (k), which is mainly related to the size, number, direction and interconnection of the rock pores.

(3) Softening

The property of rock that its strength will decrease after being immersed in water is called the softening of rock. Rock softening is related to rock porosity, mineral composition, and cementitious materials. The softening of rock is usually indicated by softening coefficient (k_d):

$$k_d = \frac{R_w}{R_d} \quad (4\text{-}6)$$

Where R_w is the ultimate compressive strength of the rock in a saturated state (kPa); R_d is the ultimate compressive strength of the rock in a dry state (kPa).

The softening coefficient of rock is less than 1. It is generally believed that when $k_d < 0.75$, the rock is called softened rock, and has weak resistance to weathering and freezing.

(4) Frost Resistance of Rock

The ability of rock to resist freezing and thawing is called frost resistance. The strength loss rate (i.e., frost resistance coefficient) and mass loss rate can be used to indicate such ability. The frost resistance coefficient refers to the ratio of the compressive strength (σ_{cd2}) of the rock after the freeze-thaw test to the compressive strength (σ_{cd1}) before the freeze-thaw test, expressed in percentage. The mass loss rate (k_m) refers to the ratio of the difference between the dry mass of the rock specimen before and after freeze-thaw test to the dry mass before freeze-thaw test, expressed in percentage. Rocks with a frost resistance coefficient of less than 75% are considered as non-frost resistant rocks.

(5) Disintegration of Rock

The disintegration of rock refers to the property of a rock that loses its cohesiveness when interacting with water and becomes a loose material that completely loses its strength. This phenomenon is caused by the weakening of the structural connections within the rock during the hydration process, and is common in sedimentary rocks cemented by soluble salts and clay. Rock disintegration is generally indicated by the rock's disintegration resistance index, which can be determined by wetting-drying cyclic tests in laboratories. For extremely soft rocks and rocks with low disintegration resistance, plasticity index, particle composition, as well as the grades of rock quality according to the disintegration of rock should be taken into consideration.

4.3 Mechanical Properties of Rock

The properties of the rock under the action of external force are called the mechanical properties of rock, including its deformation and strength characteristics. The study of the mechanical properties of rocks is mainly about their deformation characteristics, failure modes and strength.

4.3.1 Deformation of Rock

Rocks tend to deform under the action of external force, and their deformation properties could be divided into elastic, elastoplastic, plastic and brittle types. The elastic rocks could be completely restored from its deformation when removing the external force within a certain limit. However, plastic rocks cannot be completely restored from its deformation after the external force is removed, with some permanent deformations left. Brittle rocks would be destroyed without significant deformation under the action of external force. And the elastic deformation of elastic-plastic rocks tends to be less than plastic deformation before failure.

The failure states of rock specimens in the uniaxial compression tests are shown in Fig. 4-1.

Several typical uniaxial stress-strain curves of rocks could be shown in Fig. 4-2.

According to the trend of σ-ε curvature, the rock deformation and failure process can be divided into the following four stages.

(1) Compaction stage of voids and cracks (section "OA" in Fig. 4-3)

The original micro-cracks in the rock are gradually compacted under the action of the load, and the curve is concave upward. The gradient of the curve gradually increases with the increase of the load, indicating that the change of the micro-cracks starts quickly and then gradually slows down. The strength corresponding to point A is called ultimate consolidated strength. The stage is more obvious in rocks with micro-cracks, but difficult to distinguish in hard and intact rocks.

(2) The stable development stage of elastic deformation to micro-fractures (section "AB" in Fig. 4-3).

The micro-cracks in the rock are further closed and the voids are compressed. There is basically no new further development of the original cracks, and no new cracks. The stress-strain curve is approximately linear, and there is mainly elastic deformation. The stress corresponding to point B is called the elastic ultimate strength.

(3) Plastic deformation stage to failure peak stage (section "BC" in Fig. 4-3)

When the stress exceeds the elastic ultimate strength, new cracks are formed in

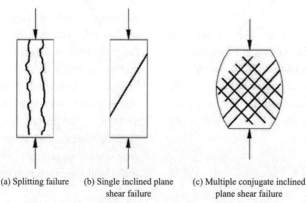

(a) Splitting failure (b) Single inclined plane shear failure (c) Multiple conjugate inclined plane shear failure

Fig. 4-1 Failure states of rock specimens in uniaxial compression tests

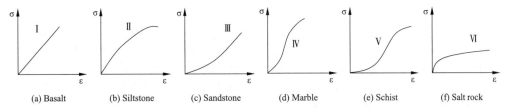

(a) Basalt (b) Siltstone (c) Sandstone (d) Marble (e) Schist (f) Salt rock

I—elastic; II—elastoplastic; III—plastic. Elastic; IV—plastic elastic-plastic; V—plastic-elastic-plastic; VI—elastic-plastic creep deformation

Fig. 4-2 Typical uniaxial stress-strain curves of rocks

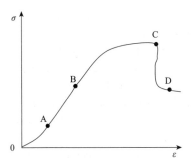

Fig. 4-3 Typical total stress-strain curve of rock

the rock, and there will be new developments in the existing cracks. The increasing rate of strain exceeds that of stress, and the gradient of the stress-strain curve gradually decreases, indicating a curvilinear relationship. The volume deformation changes from compression to expansion. With the increase of stress, the cracks continue to develop, and the rock damage zone gradually expands to form a through fracture surface, which further results in the ultimate rock failure. The stress corresponding to point C reaches the maximum value, which is called peak strength or ultimate uniaxial compressive strength.

(4) Post-failure peak falling stage to residual strength stage (after point "C" in Fig. 4-3)

After the rock is broken, the stress drops to a certain degree after a large deformation, and the stress gradually keeps stable. The stress corresponding to point D is called the residual strength.

Since most of the rock deformation has different degrees of elasticity, and the compressive stress that the building can act on the rock in engineering practice is far lower than the uniaxial ultimate compressive strength, generally, rocks could be regarded as quasi-elastic blocks to a certain extent, and its deformation characteristics can be characterized by elastic parameters, which are generally expressed by two parameters, namely the elastic modulus and the Poisson's ratio.

The elastic modulus of rock can be obtained by the ratio of the axial compressive stress to the axial strain under uniaxial compression. The greater the elastic modulus of the rock, the smaller the deformation, indicating the higher the ability of the rock to resist deformation.

The Poisson's ratio can be obtained by the ratio of lateral strain to axial strain of a rock under axial pressure. The greater the Poisson's ratio, the greater the lateral deformation of the rock.

4.3.2 Strength of Rock

The strength of rock is the ability of the rock to resist the external damage. The maximum stress that the rock can withstand before it reaches the failure is called the strength of rock. Rock failures can be crushed, broken and sheared. Hence, the corresponding strengths can be divided into the compressive strength, tensile strength and shear strength.

(1) The compressive strength of rock

Rock ultimate compressive strength (ultimate uniaxial compressive strength) is the ability of rock to resist crushing failure under uniaxial compression, which is equivalent to the maximum compressive stress when rock is damaged under compression. Formula:

$$R_c = \frac{P}{A} \quad (4\text{-}7)$$

Where R_c is the compressive strength (MPa); P is the ultimate axial force (N) when the rock fails under pressure; A is the compressive area of the rock specimen (mm²).

(2) The tensile strength of rock

The maximum tensile stress that the rock specimen can withstand under uniaxial tensile load is the tensile strength of rock, expressed by R_t. Formula:

$$R_t = \frac{P_t}{A} \quad (4\text{-}8)$$

Where R_t is the tensile strength of rock (MPa); P_t is the tensile force when the rock specimen is broken (N); A is the cross-sectional area of the specimen (mm²).

The tensile strength of rocks is far less than the compressive strength, and the tensile strengths of many rocks are less than 20 MPa.

(3) The shear strength of rock

The shear strength of rock is the ability of the rock to resist shear failure, which is equivalent to the maximum shear stress along the shear failure surface. Due to the complex mineral composition, texture and structure of rocks, there are various kinds of shear failures under normal stress, three main ones are shown in Fig. 4-4.

1) Shear-off strength of rock. Cemented rocks and rocks with solid crystalline structure tend to be sheared and fractured along a certain shear plane under normal stress, and this type of shear strength is defined as

$$\tau_f = \sigma \tan \varphi + c \quad (4\text{-}9)$$

Where c is the cohesion (kPa), φ is the internal friction angle, (°).

2) Shear strength of rock. When a rock with a structurally weak sliding surface undergoes shear failure, the ultimate shearing stress of shear failure along a weak structural plane is shear strength of rock. This kind of rock can only break along the weak structural plane, and the corresponding shear strength is usually defined as

$$\tau_f = c \quad (4\text{-}10)$$

Where c indicates the friction on the shear surface (kPa).

3) Shear-friction strength of weak plane between rocks. The shear-friction strength refers to the strength of friction between the contact surfaces of rock layers or that of rock layers and other materials under normal stress. It is usually defined as

$$\tau_f = \sigma \tan \varphi \quad (4\text{-}11)$$

Where φ is the friction angle between rocks (°).

4.3.3 Main Factors Affecting Rock Mechanical Properties

(1) Mineral composition

Rocks are composed of minerals, which has a direct impact on the physical and mechanical properties of the rock. For example, the compressive strength of quartzite is much higher than that of marble. This is because the strength of quartz is higher than that of calcite. It is worth noting that attention should be paid to factors that may reduce the strength of rocks when analyzing and evaluating the engineering geological characteristics of rocks. For example, the biotite content in granite is too high, while the clay mineral content in limestone and sandstone is too high. Biotite is

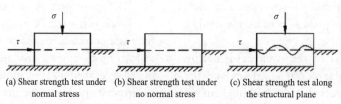

(a) Shear strength test under normal stress (b) Shear strength test under no normal stress (c) Shear strength test along the structural plane

Fig. 4-4 Three shear strength tests of rock

one of the silicate minerals with low hardness and the most developed joints. Biotite is prone to weathering and flaking, and also prone to secondary changes, eventually it can become iron oxides and clay minerals with lower strength.

(2) Texture

The textures of rocks are important factors that affect the physical and mechanical properties of rocks. According to the textures of rocks, rocks can be divided into crystalline rocks (i.e., magmatic rocks and metamorphic rocks) and cemented rocks (i.e., clastic rocks in sedimentary rocks). Compared with cemented rocks, crystalline rocks have strong binding force of crystalline particles, with small porosity, compact texture, large bulk density, small water absorption range, and higher strength. In addition, the size of mineral crystal particles also has a significant effect on strength. Generally, the smaller the particle, the higher the strength.

Cementation is the connection of mineral fragments by cements. The strength and stability of this type of rock, varying greatly, are mainly determined by the composition and form of cementation, and composition of the fragments. For example, the strength and stability of siliceous cementation is high, while that of argillaceous cementation is low, and that of iron and calcareous cementation is between that of siliceous cementation and argillaceous cementation. The common forms of cementation include substrate cementation, void cementation and contact cementation. Generally speaking, the rock porosity of the substrate cementation is small, and the strength and stability depend entirely on the composition of the cement.

(3) Structure

The influence of structure on the physical and mechanical properties of rocks is mainly determined by the uneven distribution of mineral components in the rock and the discontinuity of rock structures.

The schistose structure, slate structure, phyllitic structure, gneissic structure and rhyolite structure of the rocks often make the distribution of mineral components in the rock extremely uneven. Some minerals which are easily weathering with low strength, are mostly distributed in a certain direction, or in strips, or into local aggregates, making the physical and mechanical properties of the rock change greatly locally.

(4) water

When the rock is saturated with water, the strength of the rock will be reduced. When the rock is subjected to water, the water will infiltrate along the voids of the rock, destroying the connection between the mineral particles, and affecting the strength of rock. The effect of water on rock strength is reversible to a certain extent, and the strength can be recovered when the rock is dry.

(5) Weathering

Weathering refers to the secondary changes including gradual breaking and loosing of rocks caused by a series of physical, chemical and biological weathering when rocks are exposed to the ground surface or in the shallow underground for a long time.

4.4 Structure Types and Engineering Geological Evaluation of Rock Mass

Structures of rock mass refer to the combination and arrangement of rock blocks in the rock mass. It is mainly composed of two basic elements, i.e., structural planes and structural bodies. The structural characteristics of rock mass refer to the characteristics of the structural plane, the shape, the scale, the property and the combination relationship of the structural bodies.

4.4.1 Structural Plane

(1) Concept and classification of structural planes

Structural plane refers to a geological interface (or zone) with a certain direction, relatively low mechanical strength, and two-way extension (or a certain thickness) in the rock mass, that is, any geological interface that divides rock masses.

The structural plane of rock mass could be produced during the formation of rock mass or after its formation. Structural planes of different origins have different engineering geological characteristics. According to the cause of formation, the structural planes can be divided into three types, i.e., primary structural planes, tectonic structure planes and secondary structural planes. The geological patterns, main characteristics and engineering geological evaluation of various structural planes are shown in Table 4-1.

1) The primary structural plane is the structural plane formed during the formation of the rock, and its characteristics are closely related to the origin of the rock.

2) The tectonic structural plane is the fracture surface formed during the tectonic movement.

3) The secondary structural planes are produced under the action of external forces after the formation of rock mass.

In addition, structural planes can be divided into shear structural planes and tensile structural planes according to their stress state, for further explanation of which, please refer to the previous chapter on joints.

(2) Features of structural plane

In engineering practice, to obtain the features of structural plane is very important. The scale, shape, connectivity, filling properties and density of structural plane all have great influences on the physical and mechanical properties of structural plane. The research of the structural plane mainly includes the attitude, density, continuity, roughness, openness, filling conditions of the structural plane.

1) Attitude of structural plane

The attitude of structural plane could be expressed by the strike, dip and dip angle. Special attention should be paid to the relationship between the orientation of the structural plane and the orientation of the engineering structures (i.e., tunnels), which often plays an important role in the stability of rock mass and the safety of engineering structures.

2) Density of structural plane

The density of structural plane reflects the integrity of the rock mass, and the spacing and linear density of structural planes are usually used to express the density of structural planes.

① Linear density (K) refers to the number of structural planes per unit length (m). Generally, the linear density is the number of structural planes per meter length in the normal direction of a group of structural planes. The larger the linear density, the denser the structural plane. The values of K in different measurement directions are often different. Therefore, the values of K in the two vertical directions can reflect the anisotropy of the rock mass.

② The spacing of structural plane is often used to indicate the integrity of the rock mass in practices. The spacing of structural plane refers to the average spacing of the same group of structural planes, and it has a reciprocal relationship with the linear density of structural planes. The classification of joint spacing recommended by Chinese hydropower department is shown in Table 4-2.

3) Continuity of structural plane

Continuity refers to the distribution range of the structural plane. The relative range of the structural plane and the entire rock mass or engineering structure is the focus of continuity research. It can be described according to the

Table 4-1 Structural plane classifications and their main characteristics

Genetic types		Geological types	Main features			Engineering geological evaluation
			Attitude	Distribution	Properties	
Native structure plane	Sedimentary structural plane	1. Bedding/Bedding plane structure 2. Weak interlayer zone 3. Unconformity discontinuity 4. Sedimentary discontinuity	Generally consistent with the rock formation, it is an interlayer structural plane	The distribution of such structural planes in marine rock formations is stable, and the continental rock formations are staggered and easy to pinch	Structural planes such as bedding layers and weak interlayers are relatively flat. Unconformity surfaces and sedimentary discontinuities are mostly composed of clastic muddy materials and are uneven	Many large dam foundation slips and landslides at home and abroad are caused by such structural planes
	Igneous structural plane	1. Contact surface between intrusive body and surrounding rocks 2. Contact surface of rock vein and rock wall 3. Primary condensate joints	The dike is controlled by the structural plane, while the primary joint is controlled by the contact surface of the rock mass	The contact surface extends farther and is relatively stable, while the primary joints are often short and dense	The contact surface with the surrounding rock can have two different characteristics, i.e., fusion and failure. The primary joints are generally cracked surfaces, which are rough and uneven	Generally, it does not cause large-scale rock mass destruction, but combined with structural fractures, it can lead to rock mass slippage, such as local slippage of the dam abutment
	Metamorphic structural plane	1. Schistosity structure 2. Schist weak interlayer	The attitude is consistent with the direction of the rock formation or structure	The schistosity is short and the distribution is extremely dense. The weak interlayer of the schist extends far and has a fixed layer	The structural plane is smooth and straight, and the schism is often closed into a hidden structural plane in the deep part of the rock layer. The schist is weak interlayer, containing flaky minerals, and it is scaly	Collapses are common in shallow metamorphic sedimentary rocks, such as phyllite and other cutting slopes, and schist interlayers sometimes affect the stability of projects and underground tunnels

Continued

Genetic types	Geological types	Main features			Engineering geological evaluation
		Attitude	Distribution	Properties	
Tectonic structural plane	1. Joints (X-shape joints, Tension joints) 2. Fault 3. Interlayer dislocation 4. Feathery fissure, cleavage	Attitude has a certain relationship with the structural line, and the interlayer dislocation is consistent with the rock formation	Tensile fractures are short and small. Shear fractures extend far, and compressive fractures have a huge scale	Tensile fractures are uneven, often with secondary fillings, in a zigzag shape. Shear fractures are relatively straight, with pinnate fissures, and compressive faults have a variety of structural rocks, often containing fault gouge and mylonite	It has a great influence on the stability of rock mass. In many rock mass failure processes, tectonic structural planes can play a coordinating effect. In addition, it often causes landslides and roof fall of slopes and underground works
Secondary structural plane	1. Unloading crack 2. Weathered fissure 3. Weathered interlayer 4. Mudded interlayer 5. Secondary mud layer	Controlled by terrain and original structural plane	The distribution is often discontinuous lens, with poor extension. They mainly develop in the surface weathering zone	They are generally filled with muddy material, with poor hydraulic properties	It causes hazards on natural and artificial slopes, and sometimes also affects dam foundations, dam abutments and shallow tunnels, but it is generally cleared during construction

Grades of joint development　　　　Table 4-2

Joint spacing (m)	>2	0.5~2	0.1~0.5	<0.1
Joint development degree	Underdeveloped	More developed	developed	Extreme developed
Integrity of rock mass	Integral	Blocky	Fractured	Fragmented

traceable length of the structural plane in the outcrop, as shown in Table 4-3.

4) Roughness of structural plane

The roughness of structural plane is an important factor that determines the mechanical properties of structural plane, but its importance varies with the thickness and type of the filling. When studying the roughness of structural planes, we should first consider its undulating shape. In nature, the geometrical shapes of structural planes are very complex and can be roughly divided into five types, as shown in Table 4-4.

Generally, undulation and roughness are used to characterize the morphology of structural planes. The shape of structural plane has a great influence on the shear strength of structural plane. Generally, the straight and smooth structure plane has a lower friction angle, while the rough and undulating structure plane has a higher shear strength.

5) Openness and filling of structural plane

Openness and filling of the structural plane refers to the distance between the two walls of the structural plane, which can be divided into 3 levels, as illustrated in Table 4-5.

The mechanical properties of the closed structural plane depend on the rock properties of the two walls of structural plane and the roughness of structural plane. The shear strength of the slightly open structural plane, with the two walls of the rock often keep multiple connecting points, is greater than that of the open structural plane. For extremely

Description of Joint Continuity　　　　Table 4-3

Describe	Continuation length (m)	Describe	Continuation length (m)
Very poor continuity	<1	Good continuity	10~30
Poor continuity	1~3	Excellent continuity	>30
Medium continuity	3~10		

Morphological classification of structural planes　　　　Table 4-4

Types	Structural plane morphology	Structural plane characteristics
a	Straight	It mainly includes most bedding planes, flakes and shear fracture planes, etc.
b	Undulated	For example, wavy layers, lightly-deflected schistosity, and wavy compressive and compression-shear structural planes
c	Jagged	For example, the majority of tension-shear structural planes
d	Stepped	Step-shaped structure planes
e	Irregular	Its structural planes are bompy and uneven, such as sedimentary discontinuities, cross bedding, and secondary structural planes developing along the original fissures.

Classification by openness of structural plane　　　Table 4-5

Openness	Standard description	
<0.1 mm	Very tight	Closed
0.1~0.25 mm	Tight	Closed
0.25~0.5 mm	Partially open	Closed
0.5~2.5 mm	Open	Cracked
2.5~10 mm	Medium width	Cracked
10~100 mm	Very wide	Open
100~1000 mm	Extremely wide	Open
>1000 mm	Cavernous	Open

open and wide structural planes, the shear strength mainly depends on the material composition and thickness of the filling. Generally, when the filling material is clay, the strength is lower than that whose filling material is sandy. When the filling material is sandy, its strength is lower than that of the gravel filling material.

6) Sets of structural planes

When studying structural planes, the structural planes with similar orientations are grouped together. The sets and their number of structural planes are the main factor that determines the shape of the rock mass, together with spacing to determine the size of blocky rocks and the structure pattern of the entire rock mass.

7) Weak interlayer zone

The so-called weak interlayer zone refers to a soft rock layer interbedded in the upper and lower hard rocks, thin and easy to be softened by water, with low mechanical strength, high argillaceous or carbon content and strong extension. Compared with the surrounding rock mass, it has significantly lower strength and higher compressibility, or other unique soft characteristics. It is the weakest part of the rock mass, and a potential threat to engineering. In terms of genesis, weak interlayers can be divided into primary, tectonic and secondary weak interlayers.

The primary weak interlayer zone is a weak interlayer in nature, and formed at the same time as the surrounding rock mass. The tectonic weak interlayer zone is mainly formed by tectonic dislocation along the original weak surface or weak interlayer, or multiple times of dislocation along fracture surfaces, such as faulted fracture zones. The secondary weak interlayer zone is formed along the contact surface between rock mass, the original weak surface or the weaker interlayer, and it is formed by secondary actions (i.e., weathering and groundwater). The genetic types and basic characteristics of various weak interlayers are shown in Table 4-6.

(3) Classification of structural planes

Different types of structural planes have different scales. The larger one is like a broken zone extending tens of kilometers and tens of meters in width, while the smaller ones are joints extending tens of centimeters to tens of meters, or even tiny discontinuous cracks. Their impact on the project is different. Sometimes small structures can also control the stability of the rock mass. The engineering geological classification of structural planes could be divided into the following three categories according to their scale, as shown in Table 4-7.

Types and characteristics of weak interlayer zone

Table 4-6

Genetic types			Geological types	Basic characteristics
Primary weak interlayer zone			Sedimentary weak interlayer zone	The attitude is the same as that of the rock strata, with thin thickness and good continuity, and there is also the phenomenon of pinching. It contains many clay minerals, with thin bedding developed and low shear strength. It is easy to be weathered, mudded, softened
			Igneous weak interlayer zone	Layered or lens body, small thickness, easy to soften, low shear strength
			Metamorphic weak interlayer zone	The attitude is the same as that of rock strata, with thin thickness and poor continuity. It contains flaky minerals, which are scaly, and its shear strength is low
Tectonic weak interlayer zone			Fractured weak interlayer zone	The attitude is the same as that of rock strata, with strong continuity, and it develops along the weak interbed in the layered rock mass. The scaly material is destroyed, containing mud distributed in strips
Secondary weak interlayer zone	Weathered interlayer zone		Weathering of interlayer	The attitude is consistent with rock formation, or restricted by the attitude of rock mass, with good continuity in the weathering zone. The material is loose, broken and muddy with low shear strength
			Weathering of fracture	Developed along joints and faults, the attitude is controlled by them, and the continuity is not strong, generally limited to the vicinity of the ground surface, the material is loose, broken, muddy, and low shear strength
	Mudded interlayer zone		Interlayer mud	The attitude is the same as that of the rock strata, which is developed along the surface of the weak layer and has strong continuity, but the degree of muddy in each section is different. The weak surface is muddy and smooth, and the shear strength is low
		Secondary interlayer	Bedded plane	The attitude is restricted by rock strata with poor continuity. It develops near the surface, often in the form of a lens. The material is delicate, plastic, or even fluid, and the strength is very low
			Fracture surface	The attitude is restricted by the structural plane of the original rock, which is often steep, with poor continuity, fine material, simple structure, and poor mechanical properties

Classification of structural planes Table 4-7

Class	The characteristics of structural plane	Engineering geological significance	Representative geological structures
Class I (Faulted or filling structural planes)	Continuous or nearly continuous, with a certain extension direction. The extension length is generally greater than 100 m, and there is a influence zone with certain thickness	It could destroy the continuity of the rock mass, forming the mechanical boundary of rock mass, controlling the evolution direction, stability and calculation boundary conditions of rock mass deformation and failure	Faults, fault fracture zones, weak interlayer zones and some continuous structural planes
Grade II (Fissured or non-filled structural planes)	Approximately continuous, with a definite extension direction, and an extension length of tens of meters, which may have a certain thickness or influence zone	It destroys the continuity of the rock mass and constitutes the mechanical boundary of rock mass, which may form a certain control effect on the shear boundary of rock mass	Long and large slightly-inclined fissures, fissure concentrated zones and some continuous structural planes
Grade III (Discontinuous structure planes)	Rigid structural plane, which is short and small, randomly and intermittently distributed, extending from several meters to more than ten meters, with a statistically dominant direction	It destroys the integrity of the rock mass, making the mechanical properties of rock mass have anisotropic characteristics, affecting the deformation and failure modes of rock masses, as well as the seepage flow of rock mass	Various types of primary and structural fractures

4.4.2 Structural Body

A single rock block in a rock mass cut by a structural plane is called a structural body. Due to the combination of structural planes of various genetics, structural bodies of different sizes and shapes can be formed in the rock mass (Fig. 4-5).

Affected by the number, density, attitude, length of the structural planes, the shapes and sizes of the structural bodies in rock mass are various. However, according to the characteristics of shapes, structural bodies can be roughly summarized as column, block, slate, wedge, rhombus and cone. When the rock mass is strongly deformed and broken, flakes, fragments, and other forms of structural bodies can be formed.

The shape, size, attitude and location of structural bodies are different, and their influence on the stability of engineering rock mass is very different. When the structural body has the same shape and size, but different attitude, its stability is different in the same engineering position. When the structural body is the same in shape, size, and attitude, its stability is different in different engineering positions.

The size of structural body can be expressed by the volumetric joint count, which is defined as the total joints in the unit volume of the rock mass (the number of joints/m³). Formula:

$$J_V = \frac{1}{s_1} + \frac{1}{s_2} + \cdots\cdots + \frac{1}{s_n} = \sum_{i=1}^{n}\frac{1}{s_i} \quad (4-12)$$

Where s_i is the spacing of the i_{th} structural plane in the rock mass; $1/s_i$ is the number of joints in this group of structural planes (number of joints/m³). According to the value of J_V, the size of the structural body can be classified (Table 4-8).

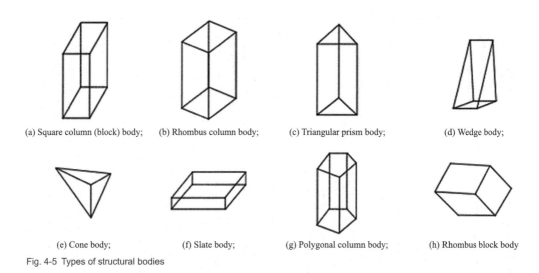

(a) Square column (block) body; (b) Rhombus column body; (c) Triangular prism body; (d) Wedge body;

(e) Cone body; (f) Slate body; (g) Polygonal column body; (h) Rhombus block body

Fig. 4-5 Types of structural bodies

4.4.3 Structure Types of Rock Mass and Their Engineering Geological Characteristics

The engineering geological characteristics of rock mass firstly depend on the characteristics of its structural planes, and secondly, the properties of structural bodies of rock mass. For example, in terms of the engineering geological characteristics, granite with loose structures is often worse than shale with stratified structure. Therefore, when analyzing the engineering geological characteristics of rock mass, firstly, we must analyze the structural characteristics of rock mass and its corresponding engineering geological characteristics, and then analyze the engineering geological characteristics of the structural bodies that make up the rock mass. Laboratory and in-situ physical and mechanical tests of rock mass is also necessary to help accurately grasp and understand the engineering geological characteristics of rock mass. The engineering geological characteristics of rock mass with different structural types are shown in Table 4-9 below.

The schematic diagram of rock mass structural patterns is as shown in Fig. 4-6.

Classification of structural bodies   Table 4-8

Description	Giant block	Large block	Medium block	Small block	Fragments
Volume number of cracks J_v (number of cracks/m³)	<1	1~3	3~10	10~30	>30

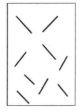 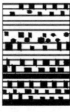  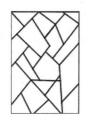 

(a) Monolithic structure; (b) Layered structure; (c) Blocky structure; (d) Fragmented structure; (e) Loose structure

Fig. 4-6 Structural patterns of rock mass

Classification of rock mass structural types and their engineering geological characteristics   Table 4-9

Structural types of rock mass	Geological types of rock mass	Main structure shapes	Structural plane development	Geotechnical characteristics	Possible geotechnical problems
Monolithic structure	Homogeneous, massive magmatic rock, metamorphic rock, huge thick sedimentary rock, orthometamorphic rock	Massive	They are mainly the primary tectonic joints, mostly closed. The distance between the fractured structural planes is greater than 1.5 m, generally no more than 1–2 groups, and there are no fall-blocks formed by the dangerous structural planes	The overall strength is high, the rock mass is stable, and it can be regarded as a homogeneous elastic isotropic body	Local sliding or collapse of unstable structures, rock bursts in deep caverns
Blocky structure	Thick layered sedimentary rock, orthometamorphic rock, massive magmatic rock, metamorphic rock	Blocky or columnar	There are only a few penetrability joints, and the spacing of the structural planes is 0.7–1.5 m, generally 2.3 groups with a few isolates	The overall strength is relatively high, the structural planes can restrain each other, and the rock mass is basically stable, similar with an elastic isotropic body	
Layered structure	Thin and medium-thick layered sedimentary rock and parametamorphic rock	Layered slate lens	Bedding, fragmentation, joints, and dislocations	Nearly uniform anisotropic body, with its deformation and strength characteristics controlled by the layer and rock layer combination, which can be regarded as an elastoplastic body with poor stability	Unstable structure bodies may cause collapse, especially flexural and tensile failures of rock masses, and the plastic deformation of weaker rocks
Fragmented structure	Fractured rock mass strongly influenced by severe structural planes	Fragmental	Faults, fault fracture zones, flakes, bedding and interlayer structural planes are relatively developed, and the distance between fractured structural planes is 0.25–0.5 m, generally in more than 3 groups, formed by many separate structural bodies	The integrity damage is severe, the overall strength is very low, and it is controlled by weak structural planes such as fractures. It is mostly an elastoplastic medium with poor stability	It is easy to cause large-scale rock mass instability, which can be aggravated by groundwater
Loose structure	Fault fracture zone, strong-weathered zone, and fully weathered zone influenced by severe structural planes	Detrital granular	The fault fracture zone intersects, and the structure and weathering fissures are dense. The structural plane and the combination are intricate, and it is filled with cohesive soil, forming many separated rocks of different sizes	The integrity is extremely damaged, with poor stability. The rock mass is close to the loose medium	It is easy to cause large-scale rock mass instability, which can be aggravated by groundwater

Quiz

1. What are the physical properties of rocks? What are the factors that affect the properties of geotechnical engineering?

2. What are the mechanical properties of rocks? What are the factors that affect the mechanical properties of rocks?

3. What is the difference and relation between rock and rock mass?

4. What is the rock mass structural planes and structural bodies? How does the structural plane affect the engineering properties of rock mass? What is the engineering geological properties of rock masses?

5. How to express the rock mass structure surface by the stereographic projection?

6. What factors determine the strength of rock mass?

Chapter 5
Classification and Engineering Geological Properties of Soils

5.1 Introduction

Soil is the most important component on the surface of the Earth's crust. It is a loose sediment formed by various complex geological processes on the surface of the lithosphere during a long geological age. The engineering geological properties of soil vary greatly due to their different formation ages, natural conditions, material composition, and structures.

5.2 Origin Types and Characteristics of Soil

Generally speaking, the Quaternary loose sediments formed in a similar geological environment are largely consistent in engineering geological characteristics. According to the division of geological origin, soil can be divided into: residual soil, deluvial soil, diluvial soil, alluvial soil, glacial soil, and aeolian soil. The main geological properties of various soils are as follows.

5.2.1 Residual Soil (Q^{el})

Residual soil is a kind of debris accumulation that remains in the place after rocks are weathered and broken, as shown in Fig. 5-1(a). The particles of residual soil have not been rounded or sorted, have no bedding structure, and have poor homogeneity, as a result, physical and mechanical properties of residual soil are quite different. At the same time, such soil particles are mostly angular coarse-grained soil with large porosity, uneven settlement can occur easily when residual soil is used as the foundation of buildings. Its water permeability is so strong that there is generally no groundwater in the residual soil. Factors influencing the engineering geological characteristics of residual soils are mainly weather conditions and conditions of the parent rocks.

5.2.2 Deluvial Soil (Q^{dl})

Deluvial soil is a slope deposit formed by the products of rock weathering being transported by the water of rain or snow to a relatively gentle slopes zone, as shown in Fig. 5-1(b). It is generally distributed at the waist or foot of the slope, and its upper part is connected with the residual soil. The newly accumulated residual soil is loose and high in compressibility. Since the lower bedrock surface is often rich in water, instability problems such as sliding along the underlying residual layer or bedrock surface are prone to occur in projects.

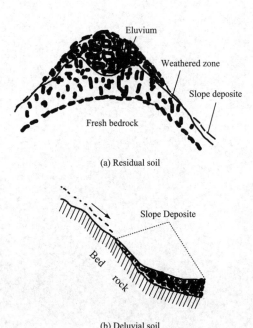

Fig. 5-1 Residual soil and deluvial soil

5.2.3 Diluvial Soil (Q^{pl})

Diluvial soil is a deposit formed in front of the mountain by the detrital material transported by heavy rains and temporary floods in mountainous areas (Fig. 5-2). Although the particles of diluvial deposits change with the distance from the mountain due to the sorting effect in the transportation process, the roundness of the particles is not good due to the short transportation distance.

(a) Diluvial fan

(b) Diluvial sediment

Fig. 5-2 Diluvial fan and diluvial sediment

5.2.4 Alluvial Soil (Q^{al})

Alluvial soil is a sediment formed by the action of river water in plain valleys or mountain valleys. It is characterized by obvious bedding structure. Due to the significant transport effect, detrital materials are sub-circular or round particles (boulders, pebbles and cobbles) that are gradually formed from angular particles (block stones, crushed stones, and breccias) through rolling and collision. The longer the transport distance, the finer the sediments. Alluvial soil can be divided into river bed facies, floodplain facies, oxbow lake facies and estuarine delta facies. Alluvial soil has different engineering geological characteristics with different formation conditions.

5.2.5 Glacial Soil (Q^{gl})

Glacial soil is a sediment formed by the action of glaciers and ice water. Generally, it can be divided into three types: moraine deposit, glacial lake deposit and glacial water deposit. The moraines are mainly accumulated near the bottom of the glacier. The particles are usually gravel with sand and clay. It is compacted by the huge pressure of the overlying ice layer, so it has high strength and are good foundations for buildings. Glacial lakes and glacial water deposits are deposits formed by glacial lakes or melted glacier water, respectively. The banded clay deposited in the glacial lake has obvious bedding, but sometimes contains a small number of boulders, forming an uneven foundation soil.

5.2.6 Aeolian Soil (Q^{aeo})

Aeolian soil is a deposit formed by wind transportation. It mainly includes loose sand and sand dunes, and the typical loess is also a kind of aeolian deposits. The characteristic of this kind of soil is that it has no bedding, and the particle size of the material deposited in the same place is very close. Aeolian loess has a loose structure, low moisture content, and collapsible after being immersed in water.

5.3 Composition and Structure of Soil

Soil is composed of solid particles, water and gas in the pores of particles. It is a multi-phase, dispersed and porous system. Generally, soil is regarded as a three-phase system, including solid phase, liquid phase and gas phase. The solid phase is also called the soil particle, which is composed of mineral particles or rock fragments of different sizes, shapes, and components, they constitute the main body of the soil. The liquid phase is the water in the pores, which partially or completely fills the intergranular pores. The gas phase refers to the air in the soil, occupying the part of the pores that are not filled by water. The three are interrelated, and define the engineering geological properties of the soil through complex physical and chemical interactions.

5.3.1 Solid Phase of Soil

(1) Particle fraction and its division

The particle size composition of soil is one of the main internal factors that determine the engineering properties of soil, therefore it is also the main basis for the classification of soil.

Soil is composed of particles of various sizes. The particle size is measured by the diameter (in mm), which is called the particle diameter (or particle size). The soil particles within a certain particle size range are called particle fraction. The relative content of particles in different particle fraction in the soil is called the particle size composition (or particle gradation), indicated by the percentage of the weight of each particle fraction to the total weight of the soil particle.

The current particle fraction classification standard is listed in Table 5-1.

The characteristics of each particle fraction are: the smaller the granule, the stronger the interaction with water. Therefore, the capillary action gradually increases from nothing to greater capillary height; the water permeability increases from large to small, even impermeable; gradually from non-viscous, non-plastic to a series of special features properties (combined with the development of water) such as greater viscosity and plasticity, and water swelling; in terms of mechanical properties, the strength gradually decreases, and the soil can be more easily deformed when subjected to external forces.

(2) Particle size analysis

The particle size and composition of the soil expressed by the relative content of the particles of each particle fraction in the soil (percentage of the total mass of the particles) is called the particle gradation of the soil. The particle gradation of the soil needs to be determined by the particle size analysis experiment of the soil. For coarse particles with a particle size greater than 0.075 mm, the soil mass of the particle fraction is determined by the sieving method. It is difficult to sieving particles with a particle size of less than 0.075 mm, and the relative content of the particle fraction can be determined by the hydrometer method or the pipette method. Cumulative curve of soil particle gradation is shown in Fig. 5-3. The gradation curve can be used to directly judge the content of each particle fraction in the soil. If the curve is steep, it means that the soil particle size is uniform and the gradation is not good; on the contrary, if the curve is flat, it means that the soil particle size is not uniform with good gradation however.

When the cumulative weight percentage of soil particles smaller than a certain particle size is 10%, the corresponding particle size is called the effective particle size d_{10}. When the cumulative weight percentage of soil particles smaller than a certain particle size is 60%, the particle size is called the limited particle size

Division of soil particle fraction   Table 5-1

Particle fraction	Particle name		Particle size range (mm)	General characteristics
Huge particles	Boulders, block stone		>200	High water permeability; no stickiness; no capillary action
	Pebbles, gravels		60<d≤200	
Coarse particles	Round gravel or breccia granules	Coarse	10<d≤20	High water permeability; no stickiness; capillary water rising height does not exceed the particle size
		Medium	5<d≤10	
		Fine	2<d≤5	
	Gravel-sand	Coarse	0.5<d≤2	Easily permeable; no stickiness; no plasticity; loose when dry; the capillary water rises little (generally less than 1 m)
		Medium	0.25<d≤0.5	
		Fine	0.1<d≤0.25	
		Whit	0.075<d≤0.1	
Fine particles	Silt	Coarse	0.01<d≤0.075	Water permeability is weak; slightly viscous when wet (capillary force connection), loose when dry, and easy to flow when saturated; no plasticity and water swellability; capillary water rises high; wet soil shakes with water precipitation (liquefaction)
		Fine	0.005<d≤0.01	
	Clay		≤0.005	Almost impervious to water; viscous and plastic when wet, swells when exposed to water, and shrinks significantly when dry; capillary water rises high, but at a slow speed

Note: The particles of boulders, pebbles and round gravel are rounded or sub-circular; the particles of block stone, gravel and breccia have edges and corners.

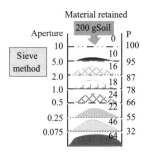 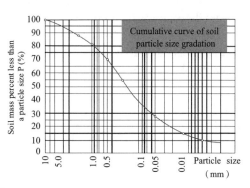

Fig. 5-3 Particle gradation curve of soil

d_{60}. The ratio of d_{10} to d_{60} reflects the degree of unevenness of particle gradation, and is called the unevenness coefficient C_u.

The larger the C_u, the more uneven the soil particles (the smoother the particle gradation accumulation curve). When such soil is used as the soil material for the filling project, a smaller pore ratio (larger density) can be easily realized. In engineering, the soil with $C_u<5$ is regarded as uniform; the soil with

$C_u > 10$ is uneven, that is, good gradation.

In addition to the coefficient of unevenness (C_u), the coefficient of curvature (C_c) can also be used to explain the curvature of the cumulative curve of soil particle gradation, so as to analyze the combined characteristics of the soil particle size components:

$$C_c = \frac{d_{30}^2}{d_{10}d_{60}} \quad (5\text{-}1)$$

In the formula, d_{30} is the particle size of soil whose cumulative content is 30%.

In engineering, the soil with $C_u \geqslant 5$ and C_c value between 1 and 3 is called well-graded soil. If these two conditions of soil cannot be met at the same time, it is called poorly graded soil. For soils with a C_c value less than 1 or greater than 3, its accumulation curve is obviously curved (concave face down or up) and is stepped, with discontinuous particle size components mainly composed of large particles and small particles, and without intermediate particles.

(3) Mineral composition of soil

The solid particles that make up the soil can be divided into four categories according to their mineral composition and their influence on the properties of soil engineering, namely: primary minerals, secondary minerals that are insoluble in water (mainly clay minerals and silicon and aluminum oxides), soluble secondary minerals and easily decomposable minerals and organic matter.

1) Primary minerals

Primary minerals are mineral fragments, with composition stay unchanged, of rock after physical weathering. The main primary minerals that make up the soil are quartz, feldspar, mica, hornblende, pyroxene, olivine, garnet and so on. The primary minerals are characterized by coarse particles, relatively stable physical and chemical properties. The primary minerals have a much smaller impact on the engineering properties of soil than other minerals.

2) Secondary minerals

New minerals with smaller particles formed from primary minerals that are further decomposed by weathering. Secondary minerals are divided into two types: one is soluble secondary minerals formed by the precipitation of soluble substances in primary minerals that are filtered and carried by water; the other is insoluble secondary minerals, which is the residue after the soluble part of the primary mineral is leached, and composition and structure of the primary mineral are changed.

The secondary minerals that are insoluble in water mainly include: ① Clay minerals, which is hydrous aluminosilicate, mainly including three basic categories: kaolinite, illite and montmorillonite. ② Secondary SiO_2: including colloidal and quasi-colloidal SiO_2. ③ Sesquioxide: including minerals composed of trivalent Fe, Al, O, OH, and H_2O. They are the main components of clay particles.

However, different types of these minerals have different effects on the engineering properties of soil. Only in terms of each category of clay minerals, the impact is also significantly different. The reason is that they have different chemical compositions and crystal lattice structures.

① Kaolinite: Each unit cell of the crystalline framework of kaolinite is composed of an aluminum hydroxide octahedral layer and a silicon oxygen tetrahedral layer, which is a 1:1 type (or two-story) structural unit layer, as shown in Fig. 5-4.

② Montmorillonite: The crystal lattice of montmorillonite minerals is different from that of kaolinite minerals. Its cell is composed of two silica tetrahedron layers with one aluminum hydroxide octahedron layer in the middle, which is 2:1 type (or three-layer type) structural unit layer, as shown in Fig. 5-5.

③ Illite: Its crystal cells belong to the unit layer of 2:1 type structure as montmorillonite. The difference is that some ions in the silica

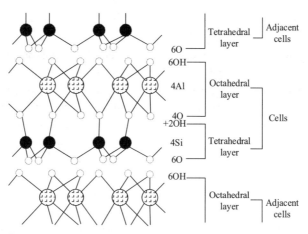

Fig. 5-4 Crystal structure of kaolinite

tetrahedron are often replaced by Al and Fe. Therefore, a number of monovalent positive ions will appear between adjacent crystal cells to compensate for missing positive charge in the crystal cells, and connect adjacent crystal cells, as shown in Fig. 5-5 and Fig. 5-6.

Colloid activity, hydrophilicity and engineering properties of secondary Si, sesquioxide and other minerals in soil are generally smaller than those of clay minerals.

(4) Soluble secondary minerals

Soluble secondary minerals are also called water-soluble salts. According to their solubility in water, they are divided into three types: soluble salts, moderately soluble salts and insoluble salts. When there is little water in the soil and the secondary mineral crystalline soil contains a certain amount of water-soluble salt, mineral crystals will precipitate, and this type of salt will cement in the soil; when the soil has a lot of water, the salt will dissolve and the connection of the soil will be broken. Therefore, when the soil contains a certain amount of water-soluble salt, the properties of the soil will change greatly with the crystallization or dissolution of minerals, especially soluble salt and moderately soluble salt. They are harmful components in the soil, and engineering construction has certain requirements for their content.

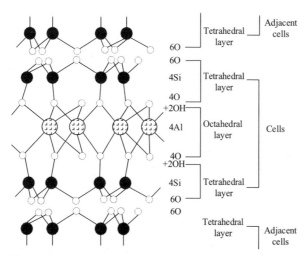

Fig. 5-5 Crystallization structure of montmorillonite

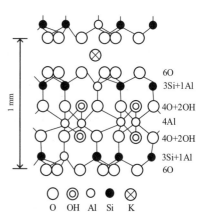

Fig. 5-6 Crystal structure of illite

(5) Organic matter

Organic matter in soil is generally a mixture that is firmly combined with other components constituting soil particles. The effect of organic matter on the engineering properties of soil is that it has stronger colloid properties and higher hydrophilicity than clay minerals. Therefore, the influence of organic matter on soil properties is more intense than clay minerals. The influence of organic matter on soil engineering properties mainly depends on the following factors:

1) The higher the organic matter content, the greater the impact on soil properties.

2) Soil saturation is different, the influence of organic matter on soil properties is different.

When the soil containing organic matter is dry, the organic matter can play a strong connection role between particles. When the soil moisture content increases, the organic matter will make the water supply and hydration film of soil particles thicken sharply, weaken the connection between soil particles, and significantly reduce the strength of soil.

3) The higher the degree of decomposition of organic matter, the greater the impact.

4) It is related to the thickness, distribution uniformity and distribution mode of organic matter in soil.

5.3.2 Liquid Phase of Soil

Water exists in soil in different forms and states. Different types of water have different effects on the engineering geological properties of soil. According to the position of water molecules in soil, it can be divided into mineral water and water in pores. The pore water can be classified as bound water and free water (non-bound water), while bound water includes strong bound water and loosely bound water.

Water existing in the crystal lattice of soil particles or participating in the composition of mineral lattice is called crystal water or bound water within mineral. It can be converted into gaseous water and separated from soil particles only at high temperature (140~700℃). Therefore, from the perspective of the influence on the engineering properties of soil, the internal bound water and crystalline water in minerals should be regarded as part of mineral particles.

(1) Bound water

Bound water refers to the soil water attached to the surface of soil particles by molecular gravity and electrostatic attraction. This attraction is as high as thousands to tens of thousands of atmospheric pressures, so that water molecules and soil surface can be firmly bonded together. Bound water cannot transfer hydrostatic pressure, cannot flow arbitrarily, and its freezing point is below zero.

Since the surface of soil particles is generally negatively charged and forms an electric field, water molecules and cations (such as Na^+, Ca^{2+}, Al^{3+}, etc.) in aqueous solution within the electric field range of soil particles are easily adsorbed on the surface of soil particles. Because the water molecule is a polar molecule, it is attracted by the surface charge of soil particles or the ionic charge in aqueous solution and aligned (Fig. 5-7).

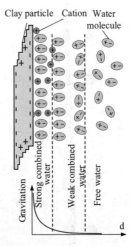

Fig. 5-7 The change of electric double layer, combined water and electrostatic attraction on soil surface

Because the distance between bound water and particle surface are different, the electric field force on bound water is different, consequently bound water can be divided into strong bound water and loosely bound water.

The electrostatic attraction strength of the soil particles decreases with the increase of the distance from the surface of the soil particles. The water molecules near the surface of the soil particles are strongly attracted by the soil particles (up to 1000~2000MPa), and lose the ability of free movement. This type of water is called strong bound water or suction water.

Water molecules away from the surface of soil particles are under smaller attraction of soil particles, therefor, they have some freedom of being active, and are loose and irregular in arrangement, this type of the water is called loosely bound water or thin film water. Loosely bound water has much larger thickness than strong bound water, and changes greatly. It is the main body of the whole bound water film. It still cannot transmit hydrostatic pressure, has no solubility, and its freezing point is lower than 0℃.

(2) Free water

Free water refers to the general liquid water in soil pores beyond the scope of electrostatic attraction on the surface of soil particles. Mainly controlled by gravity, it can transfer hydrostatic pressure and dissolve salt, and freeze to ice at about 0℃. The transition type between gravitational water and bound water is capillary water.

1) Liquid water

Capillary water is a transitional type of water that exists above the underground water level in the small pores of soil due to the capillary force combined with the molecular gravity of soil particles and the surface tension of water and air interface. Capillary water can be divided into capillary ascending water and capillary suspended water according to its location and the relationship with the underground water surface formed by gravity water. Gravity water is a kind of water that exists in relatively coarse pores and has the ability of free movement. It flows under the action of gravity and is ordinary liquid water. When gravity water flows, hydrodynamic pressure is generated, which can scour the fine soil particles in the soil. This effect is called mechanical subsurface erosion. Gravity water can also dissolve water-soluble salt in soil, which is called chemical potential erosion. Both of them will increase the porosity of soil, increase compressibility and reduce shear strength of soil. At the same time, the weight of saturated soil below the underground water table and the weight of engineering structure will be relatively reduced due to the effect of gravity water buoyancy.

2) Solid water

Solid water is ice, which exists in soil in the form of ice interlayer, ice lens and fine ice crystals. And the soil is cemented to form frozen soil with its strength boosted. However, the strength of soil after thawing is often lower than that before freezing.

3) Gaseous water

Gaseous water exists in the state of water vapor, which strictly belongs to the gas phase of soil. Under the pressure difference, the gaseous water moves from the place with high pressure to the place with low pressure. The gaseous water and liquid water in soil pores maintain a certain dynamic equilibrium under certain temperature and pressure conditions. When the pressure remains constant and the temperature rises, some liquid water evaporates to gaseous water. When the temperature decreases, some gaseous water condenses into liquid water.

5.3.3 Gas Phase of Soil

Gas is also one of the components of soil. O_2, CO_2 and N_2, are major gas components in soil, together CH_4 and H_2S, which are basically consistent with the composition of the atmosphere. However, the relative content of various gases in the soil is quite different from that in the atmosphere. The adsorption intensity of soil particles to gases is $CO_2 > N_2 > O_2 > H_2$, which leads to the relative content of CO_2 in the soil up to 10%, while CO_2 in the atmosphere only accounts for 0.03%.

Gas in soil pores exists in two forms, one is closed gas, and the other free gas. Free gas usually exists in the unsaturated zone near the surface and is connected to the atmosphere. It has exchange effect with the atmosphere with the change of external conditions and is in a dynamic equilibrium state. Free gas content depends on the volume of soil pores and the filling degree of water. It generally has little effect on the properties of soil. The closed gas exists in the soil pores in a closed state, which is formed in the process underground water level rising where some gas cannot escape due to different and complicated pore size of soil and is surrounded by water. It is isolated from the atmosphere and exists in some pores in a closed state. Closed gas can have a great influence on the properties of soil.

5.3.4 Structure of Soil

The structure of soil refers to the comprehensive characteristics of the size, shape, surface characteristics, connection relationship and arrangement of soil particles.

(1) Types of soil structure

The soil structure can be divided into two basic types according to its particle arrangement and connection: single-grain (granular) structure and aggregate (agglomerate) structure. The formation and change of these two different structural characteristics depend on the particle composition, mineral composition and environmental conditions of the soil.

1) Single-grain

Giant-grain soil and coarse-grain soil usually have single-grain structure stacked by single particles (Fig. 5-8). This structure is formed by the accumulation of soil particles due to gravity. Giant-grain and coarse-grained soils have larger particles. The molecular gravity between soil particles is relatively small, there is almost no connection between particles, or the connection is very weak. The properties of giant-grain and coarse-grained soils depend mainly on the size and arrangement of soil particles.

In giant grained soil and gravel soil, due to the different content of coarse and fine particles, they have different structural forms. If the content of coarse particles is high, they are directly contacted with each other, and fine particles are filled in their pores, which is called coarse stone structure (Fig. 5-9a). If the content of coarse matter is low, it is surrounded by fine matter and cannot be directly contacted, which is called pseudo-spot structure (Fig. 5-9b).

2) Aggregate structure

Fine-grained soil particles are small and colloid, and cannot condense as a single particle in water and deposit by forming a more complex aggregate and form a unique aggregate structure of fine-grained soil.

The aggregate structure of fine-grained soil can be divided into two types according to whether the soil particles are uniform or not: homogeneous particle structure and heterogeneous particle structure. Homogeneous

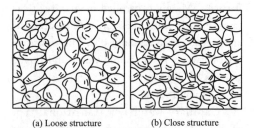

(a) Loose structure   (b) Close structure

Fig. 5-8 Single grain structure of soil

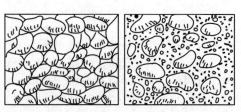

(a) Coarse stone structure   (b) Pseudo-spot structure

Fig. 5-9 The structure of giant grained soil and gravel soil

aggregate structure is divided into honeycomb and flocculent structure. The honeycomb structure is formed when 0.02 ~0.002 mm soil particles sink in water and through the connect together (Fig. 5-10a). The flocculent structure is formed by condensation of soil water with particle size less than 0.002 m (Fig. 5-10b). The heterogeneous aggregate structure is formed by the structure filled with clay aggregate between silt and gravel (Fig. 5-10c).

Aggregate structure is unstable, vulnerable to external influence and change, such as compression, shear, heating, drying, wetting and other effects as well as the change of ion composition, concentration, pH value of aqueous solution, which can make the structure change and cause the change of soil properties. Soil with aggregated structure has the following characteristics:

① The porosity is large (up to 50%~98%), and the diameter of each individual pore is small, the pores of the aggregate flocculent structure are even smaller with larger porosity. Therefore, the compressibility of soil is larger.

② Water capacity, moisture content is very large, often more than 50%, and because of the combination of water, such soil has difficulties in drainage, so its compression process is slow.

③ Variability-Instability is large. It is sensitive to changes in external conditions (such as pressure, vibration, drying, wetting, and changes in composition and properties of aqueous solution), and often suffers qualitative changes. So aggregated structure is also called variable structure. The thixotropy of soft clay, for example, is a special property resulting from the instability of such structures

(2) Tectonics of soil

In a certain soil, tectonics of soil refers to the shape and comprehensive characteristics of soil elements with relatively uniform structure, and the size, shape, arrangement and relationship of soil unit. The interface of a unit is called a structural plane or level. The shape of the unit is mostly layered, banded and lenticular, and the interface morphology is flat and wavy. The size of the unit is usually represented by thickness and extension length.

Tectonics of soil is formed by complex interaction with various factors in the process of soil formation and change. So, every genetic type of soil has its unique tectonics. Common tectonics of soil are as follows:

1) Layered tectonics

Soil layer is bedding composed of different color and different particle size soil, and is usually horizontal in the plain area. Layered tectonics is an important feature of fine-grained soil.

2) Dispersion tectonics

The distribution of soil particles in the soil layer is uniform and the properties are similar. For example, both sand and pebble layers have dispersed tectonics.

3) Combined tectonics

There are coarse particles or various nodules in fine-grained soil, such as silty clay containing stone, and moraine soil containing gravel. Its engineering properties depend on the part of fine-grained soil.

4) Fracture tectonics

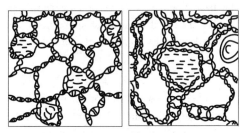

(a) Honeycomb structure   (b) Flocculent structure

(c) Heterogeneous aggregate structure

Fig. 5-10 Fine-grained soil aggregate structure diagram

There are many discontinuous small cracks in the soil, and some hard plastic and hard clay have such tectonics. Such tectonics is characterized by low fracture strength, high permeability and poor engineering properties, which can greatly reduce the strength and stability of soil, increases the permeability, and unfavorable to engineering construction.

5.4 Physical and Mechanical Properties of Soil

5.4.1 Three-phase Ratio Index of Soil

The index that represents the three-phase proportional relationship of soil is called the three-phase proportional index of soil, which is also called the basic physical properties of soil, including particle proportion, weight, moisture content, saturation, pore ratio and porosity of soil. The three-phase indicator diagram of soil is shown in Fig. 5-11.

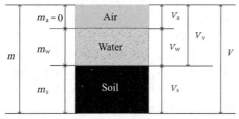

Fig. 5-11 Soil three-phase indicator diagram

The meaning of the symbols in the figure is as follows: m_s is soil weight, m_w is the weight of water in soil, m is the total weight of soil, $m = m_s + m_w$; V_s is volume of soil particle, V_w is the volume of water in soil, $\gamma_w \cdot V_w = m_w$ ($\gamma_w = 1 \text{g/cm}^2 \approx 10 \text{kN/m}^2$), V_a is the volume of air in soil, V_v is the volume of pore in soil. $V_v = V_a + V_w$; V is the total volume of soil,

$V = V_a + V_s + V_w$.

(1) Soil particle density and proportion

Soil particle density refers to the ratio of solid particle mass m_s to its volume V_s, that is, the mass of soil particles per unit volume (g/cm³), formula:

$$\rho_s = \frac{m_s}{V_s} \qquad (5\text{-}2)$$

The density of soil particles, generally between 2.60~2.80 g/cm³ (Table 5-2), is only related to the mineral composition of soil particles, and has nothing to do with the porosity and moisture content of soil. In general case, the density of soil particles will decrease with the increase of organic matter content, and increase with the increase of mafic minerals. Soil particle density is the actual measured index.

The ratio of the weight of soil particles to the mass of pure water at 4℃ of the same volume is called the specific gravity of soil particles (the relative density of soil particles d_s), formula:

$$d_s = \frac{m_s}{V_s} \cdot \frac{1}{\rho_{w_1}} \qquad (5\text{-}3)$$

Where ρ_{w_1} is the weight per unit volume of water at 4℃, usually 1 g/cm³. In fact, the specific gravity of soil particles is numerically equal to the density of soil particles, which can be measured in a specific gravity bottle in the

	Soil particle density of main types				Table 5-2
The type of soil		Gravelly soil	Sandy soil	Silt	Silty clay	Clay
Soil particle density (g/cm³)	Common value	2.65~2.75	2.65~2.70	2.65~2.70	2.68~2.73	2.72~2.76
	Average value	—	2.66	2.68	2.71	2.74

laboratory.

(2) Soil density

The density of soil refers to the ratio of total mass to total volume of soil, namely the mass of soil per unit volume (g/cm^3). According to the different states of soil, the density of soil can be divided into natural density, dry density and saturated density.

1) Natural density

The mass of unit volume soil in natural state is called natural density:

$$\rho=\frac{m}{V}=\frac{m_s+m_w}{V_s+V_v} \quad (5\text{-}4)$$

In the formula, ρ is the natural density of soil (g/cm^3); m is the total mass of soil (g); V is the total volume of soil (cm^3); m_s is the mass of solid particles in soil (g); m_w is the quality of water in the soil (g); V_v is the pore volume in soil (cm^3).

The natural density, generally 1.6~2.2 g/cm^3 (Table 5-3), depends on the mineral composition, porosity and moisture content of the soil. Natural density is smaller than soil particle density in numerical value, which is an actual measured index and can be directly measured indoor and outdoor, and is an indispensable parameter in engineering geological calculation.

Average natural density of several common soils
Table 5-3

Name	Sandy	Silt	Silty clay	Clay
Natural destiny (g/cm^3)	1.40	1.60	1.60~1.75	1.80~2.00

2) Dry density

The density of soil when there is no water in the pores of soil is called dry density of soil (ρ_d), this means the quality of dry soil per unit volume, formula:

$$\rho_d=\frac{m_s}{V} \quad (5\text{-}5)$$

Dry density has nothing to do with moisture content in soil, only depends on the mineral composition and porosity of soil. The dry density can be measured and converted by other indicators. The dry density of soil is generally in the range of 1.4~1.7 g/cm^3.

3) Saturated density

The density of soil when soil pores are completely filled with water is called saturated density, which refers to the mass of unit volume soil when all soil pores are filled with liquid water, formula:

$$\rho_{sat}=\frac{m_s+V_v\cdot\rho_w}{V} \quad (5\text{-}6)$$

In the formula: ρ_w is the density of water (g/cm^3).

The saturated density of soil is the maximum of the density of soil, is a calculation index, the common value is 1.80~2.30 g/cm^3.

4) Floating density (ρ')

The floating density of soil is the difference between the mass of soil particles in unit volume and the mass of water in the same volume, formula:

$$\rho'=\frac{m_s-V_s\cdot\rho_\omega}{V} \quad (5\text{-}7)$$

(3) Gravity, dry gravity, saturated gravity and floating gravity of soil

The weight of soil per unit volume is most called the gravity of soil (kN/m^3), formula:

$$\gamma=\frac{W}{V}=\frac{W_s+W_w}{V_s+V_v} \quad (5\text{-}8)$$

The gravity of soil depends on the weight of soil particles, the size of pore volume and the weight of water in pores, and the gravity comprehensively reflects the composition and structural characteristics of soil.

The weight of solid particles in unit volume of soil is called the dry gravity γ_d of soil, formula:

$$\gamma_d=\frac{W_s}{V} \quad (5\text{-}9)$$

Dry gravity is often used as a standard to evaluate the compactness of soil in engineering to control the construction quality of filling projects.

The unit volume gravity of soil filled with water is called the saturated gravity of soil γ_{sat}, formula:

$$\gamma_{sat} = \frac{W_s + V_v \cdot \gamma_w}{V} \quad (5\text{-}10)$$

The effective weight of soil particles in unit soil volume, that is the weight of soil particles in unit soil volume subtracts the buoyancy under the groundwater level, is the floating gravity or underwater gravity γ' of soil, formula:

$$\gamma' = \frac{W_s - V_s \cdot \gamma_w}{V} \quad (5\text{-}11)$$

(4) Moisture content of soil

There are great differences in the degree of dryness and wetness of soil in nature, which can be expressed by two water-containing indexes, namely moisture content and degree of saturation.

1) Moisture content

The ratio of the mass of water in soil to the mass of solid particles is called moisture content, which is usually expressed by percentage.

$$\omega = \frac{m_w}{m_s} \times 100\% \quad (5\text{-}12)$$

For the soil with the same structure, the greater the moisture content, the more water in the soil. Due to the different natural conditions and soil porosity, soil moisture content varies greatly. Moisture content is a measured index and an important data for calculating dry density, porosity and saturation, which is usually measured by drying method.

The moisture content in soil when it pores are filled with liquid water, called saturated moisture content ω_{sat}:

$$\omega_{sat} = \frac{V_v \rho_w}{m_s} \times 100\% \quad (5\text{-}13)$$

Saturated moisture content can not only reflect the moisture content characteristics of soil pores filled with ordinary liquid water, but also the porosity of soil. Because for a certain kind of soil, the quality of soil particles is constant, with the increase of pore volume, saturated moisture content increases. Therefore, saturated moisture content is an indicator of soil porosity characterized by mass ratio.

2) Degree of saturation

Moisture content is an absolute index, which can only indicate the content of water in soil, but cannot reflect the degree of water filling in pores. Soil saturation is the percentage of water volume to pore volume in soil, indicating the filling degree of water in pores, formula:

$$S_r = \frac{V_w}{V_v} \times 100\% \quad (5\text{-}14)$$

Degree of saturation is a calculation index. The saturation varies from 0% to 100%: when the soil is dry, the saturation is zero. When all soil pores are filled with water, the saturation is 100%. In engineering research, natural clay with saturation greater than 95% is usually regarded as fully saturated soil. When the sand saturation is greater than 80%, it can be considered to have reached saturation.

(5) Pore ratio and porosity of soil

The size, quantity and connectivity of pores in soil are called porosity of soil. The number of pores in soil is usually expressed by porosity and pore ratio. They can only reflect the total volume of soil pores, rather than the size of a single pore, mainly used to indicate the degree of soil compactness. Porosity is the ratio of the total volume of pores in soil to the total volume of soil, also known as degree of porosity, commonly indicated by percentage.

$$n = \frac{V_v}{V} \times 100\% \quad (5\text{-}15)$$

Pore ratio is the ratio of the total volume of pores in soil to the total volume of solid particles in soil, which is commonly expressed by decimals.

$$e = \frac{V_v}{V_s} \quad (5\text{-}16)$$

Both porosity and pore ratio are indicators reflecting soil porosity, and the relationship between the two indicators:

$$n = \frac{e}{1+e} \times 100\% \quad (5\text{-}17)$$

$$e = \frac{n}{1-n} \quad (5\text{-}18)$$

The porosity and pore ratio of soil mainly depend on the particle size, composition and structure of soil. Porosity and pore ratio indicate the relative value of pore volume in soil. The porosity and pore ratio of natural soil depend on the structural state of soil, so they are important indicators of structural characteristics of soil. The larger the value, the larger the pore volume and the looser the soil structure; conversely, the smaller the value, the denser the structure.

Pore ratio shows the compactness of soil. According to the pore ratio, the sand is divided into four dense degree, namely dense, medium dense, slightly dense and loose (Table 5-4). It is an essential index to study the compressibility of soil and an important index to determine the basic value of bearing capacity of fine-grained soil foundation.

(6) Conversion relationship of basic physical properties of soil

Among the three-phase ratio indexes mentioned above, the relative density d_s, moisture content ω and gravity γ of soil particles can be obtained by experiments. After measuring these three basic indicators, the remaining indicators can be derived. The conversion relationship is shown in Table 5-5.

5.4.2 Physical State Index of Non-cohesive Soil

Non-cohesive soil generally refers to gravel soil and sandy soil, and silty soil is the transition type of sand and clay soil. However, silty soil is listed in non-cohesive soil for further study as its the material composition, structure and physical and mechanical properties are close to those of sand (especially sandy silty soil).

The so-called physical state refers to the compactness of soil for non-cohesive soil, the soft and hard degree of soil or soil consistency for cohesive soil. The compactness can be determined by the method of determining the relative compactness by indoor test pore ratio and in-situ test such as standard penetration test. Relative compactness is represented by the ratio D_r of the difference between the maximum pore ratio e_{max} and the natural pore ratio e, and the difference between the maximum pore ratio e_{max} and the minimum pore ratio e_{min}, formula:

$$D_r = \frac{e_{max} - e}{e_{max} - e_{min}} \quad (5\text{-}19)$$

In the formula: e_{max} is the maximum pore ratio, that is, the pore ratio under the loosest state. e_{min} is the minimum porosity ratio, that is, the porosity ratio in the most compact state. e is pore ratio. The compactness of sand can be classified by relative density (Table 5-6).

The China's standard *Code for Investigation of Geotechnical Engineering* (GB 50021—2001) stipulates the standard for dividing the tight state of sand according to the standard penetration hammer number N (Table 5-7).

Classification of sand by pore ratio Table 5-4

Name	Density			
	Dense	Medium dense	Slightly dense	Loose
Gravel sand, coarse sand, medium sand	$e<0.6$	$0.60 \leqslant e \leqslant 0.75$	$0.75 \leqslant e \leqslant 0.85$	$e>0.85$
Fine sand, silt	$e<0.70$	$0.70 \leqslant e \leqslant 0.85$	$0.85 \leqslant e \leqslant 0.95$	$e>0.95$

Basic physical properties of soil index conversion   Table 5-5

	$w(\%)$	$\rho(\text{g/cm}^3)$	G_s	$\rho_d(\text{g/cm}^3)$	e	$n(\%)$	$S_r(\%)$
w		$\dfrac{100\rho}{\rho_d} - 100$	$\dfrac{eS_r}{G_s}$	$\dfrac{S_r(G_s-\rho_d)}{G_s\rho_d}$	$\dfrac{100\rho(1+e)}{G_s} - 100$	$\dfrac{nS_r}{G_s(100-n)}$	$\dfrac{eS_r}{(1+e)\rho_d}$
ρ	$(1+0.01w)\rho_d$		$\dfrac{G_s(1+0.01w)}{1+e}$	$\rho_d + \dfrac{0.01S_r \cdot e}{1+e}$	$\dfrac{eS_r(1+0.01w)}{(1+e)w}$	$\dfrac{nS_r(1+0.01w)}{100w}$	$\dfrac{0.01nS_r+(100-n)G_s}{100}$
G_s	$\dfrac{\rho(1+e)}{1+0.01w}$	$\dfrac{S_r\rho}{S_r(1+0.01w)-w\rho}$		$\dfrac{100\rho_d}{100-n}$	$\rho_d(1+e)$	$\dfrac{100\rho-0.01S_r \cdot n}{100-n}$	$\rho(1+e)\cdot 0.01S_r \cdot e$
ρ_d	$\dfrac{nS_r}{100w}$	$\dfrac{\rho}{1+0.01w}$	$\dfrac{S_rG_s}{wG_s+S_r}$		$\dfrac{G_s}{1+e}$	$G_s(1-0.01n)$	$\dfrac{eS_r}{(1+e)w}$
e	$\dfrac{G_s(1+0.01w)}{\rho} - 1$	$\dfrac{w\rho}{S_r(1+0.01w)-w\rho}$	$\dfrac{G_s-\rho}{\rho-0.01S_r}$	$\dfrac{G_s}{\rho_d} - 1$		$\dfrac{n}{100-n}$	$\dfrac{w\rho_d}{S_r-w\rho_d}$
n	$100 - \dfrac{100\rho}{G_s(1+0.01w)}$	$\dfrac{100w\rho}{S_r(1+0.01w)}$	$\dfrac{100(G_s-\rho)}{G_s-0.01S_r}$	$100 - \dfrac{100\rho_d}{G_s}$	$\dfrac{100e}{1+e}$		$\dfrac{100w\rho_d}{S_r}$

Classification of sand by relative density   Table 5-6

Compactness	Loose	Medium dense	Dense
Relative density	$0 < D_r \leq 0.33$	$0.33 < D_r \leq 0.67$	$0.67 < D_r \leq 1$

Determination of sand density by standard hammer number N Table 5-7

Compactness	Loose	Slightly dense	Medium dense	Dense
N	$N \leq 10$	$10 < N \leq 15$	$15 < N \leq 30$	$N > 30$

5.4.3 Physical State Index of Cohesive Soil

(1) Limit moisture content

With the change of moisture content, the cohesive soil changes from a consistency state to another consistency state. The moisture content of the corresponding transition point is called the limit moisture content, also known as the consistency limit. Limit moisture content is an important characteristic index of cohesive soil, which is of great significance for the evaluation and classification of engineering properties of cohesive soil. Moreover, various cohesive soils have different limit moisture content. Among the consistency limits, the plastic limit (w_p) and liquid limit (w_L) are the most significant. The limit moisture content from plastic state to flowing state is called liquid limit w_L (also called plastic limit or flow limit). The limit moisture content of soil from semi-solid to plastic state is called the plastic limit w_p (also known as the plastic limit), which is expressed by percentage. At present, cone liquid limit instrument is generally used to determine the liquid limit of cohesive soil in China. The plastic limit of cohesive soil is generally determined by the rubbing method. The detailed determination method can be referred to the relevant soil mechanics books or *Standard for Geotechnical Testing Method* (GB/T 50123—2019).

(2) Plasticity index and liquidity index

The plasticity of cohesive soil is expressed by plasticity index. Plasticity index is the difference between liquid limit and plastic limit of cohesive soil, that is, the variation range of moisture content still in the plastic state. It is expressed by I_p, and the sign of % of moisture content is usually omitted in application.

$$I_p = w_L - w_p \quad (5\text{-}20)$$

The larger the plastic property, the larger the variation range of moisture content of cohesive soil in plastic state, namely, the stronger the plasticity of soil. Conversely, the weaker the plasticity.

Since the plasticity index comprehensively reflects various factors affecting the characteristics of fine-grained soil to a certain extent, the cohesive soil can be classified according to the plasticity index. Table 5-8 is Chinese classification standard of clay in *Code for Design of Building Foundation* (GB 50007—2011).

Classification of cohesive soil　Table 5-8

Plasticity index I_p	Name
$I_p > 17$	Clay
$10 < I_p \leqslant 17$	Silty clay

Liquidity index is the ratio of the difference between the natural moisture content and plastic limit of soil, and the difference between liquid limit and plastic limit, which is expressed by decimal, formula:

$$I_L = \frac{w - w_p}{w_L - w_p} = \frac{w - w_p}{I_p} \quad (5\text{-}21)$$

Liquidity index characterizes the consistency of fine-grained soil in natural state. When the natural moisture content of soil is less than or equal to the plastic limit, $I_L \leqslant 0$, the soil is in a hard state. When the natural moisture content is greater than the liquid limit, $I_L > 1$, the soil is in a flow state. When the natural moisture content is between liquid limit and plastic limit, $0 < I_L \leqslant 1$, the soil is in plastic state. Table 5-9 show the standard for dividing the consistency state of fine-grained

Classification of fine-grained soil consistency by liquidity index value　Table 5-9

Liquidity index	$I_L \leqslant 0$	$0 < I_L \leqslant 0.25$	$0.25 < I_L \leqslant 0.75$	$0.75 < I_L \leqslant 1.00$	$I_L > 1.00$
Consistency state	Rigid	Rigid plastic	Plastic	Soft-plastic	Flow plastic

soil by liquidity index in *Code for Design of Building Foundation* (GB 50007—2011).

5.4.4 Mechanical Properties of Soil

The mechanical properties of soil refer to the properties of soil under external forces, mainly including the compressibility of volume reduction under compressive stress and the shear resistance to shear failure under shear stress, as well as the properties under dynamic loading.

(1) Compressibility and its indicators of soil

The characteristics of soil volume reduction under pressure is called soil compressibility. The indexes of soil compression deformation are: compression coefficient (a), compression index (C_c), lateral expansion coefficient (Poisson's ratio μ), compression modulus (E_s), etc.

The compressibility of soil can be used to draw the relationship curve between pore ratio e and pressure p, which is called compression curve (Fig. 5-12).

On the compression curve, when the variation range of pressure is not large, the corresponding segment M_1M_2 on the compression curve can be approximately replaced by a straight line. If the pressure at point M_1 is p_1, the corresponding pore ratio is e_1. The pressure at point M_2 is p_2, and the corresponding pore ratio is e_2. The slope of the M_1M_2 segment can be expressed as the following formula:

$$a = \tan\alpha = \frac{e_1 - e_2}{p_2 - p_1} \quad (5-22)$$

In the formula, a is the compression coefficient. At the same time, the above equation shows that the compression coefficient of the same soil is not a constant, but varies with the variation range of pressure.

In order to facilitate the application and comparison, and take the pressure variation range of general building foundation into consideration, the compression coefficient a_{1-2} obtained when the pressure interval increases from $p_1 = 0.1$ MPa to $p_2 = 0.2$ MPa is generally used to evaluate the compressibility of soil.

Low compressibility soil: $a_{1-2} < 0.1$ MPa^{-1}

Medium compressibility soil: 0.1 MPa$^{-1} \leq a_{1-2} < 0.5$ MPa^{-1}

High compressibility soil: $a_{1-2} \geq 0.5$ MPa^{-1}

According to the compression test data, another commonly used compressibility index-compression modulus E_s can be obtained. Compressive modulus refers to the ratio of the stress σ_z in the compression direction to the corresponding strain ε_z when the soil is subjected to compression under lateral confinement, formula:

$$E_s = \frac{\sigma_z}{\varepsilon_z} \quad (5-23)$$

In the formula: $\sigma_z = p_2 - p_1$, $\varepsilon_z = \frac{\Delta h}{h} = \frac{e_1 - e_2}{1 + e_1}$.

Both σ_z and ε_z can be obtained from compression tests. By substituting the two formulas into the above formula, it is concluded that:

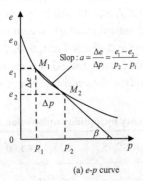

(a) e-p curve

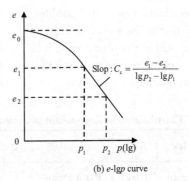

(b) e-lg$p$ curve

Fig. 5-12 Compression curve of soil

$$E_s = \frac{p_2 - p_1}{e_1 - e_2}(1 + e_1) = \frac{1 + e_1}{a} \quad (5\text{-}24)$$

In the formula: a is the compression coefficient when the pressure increases from p_1 to p_2. e_1 is the pore ratio at pressure p_1.

In engineering practice, the compression modulus corresponding to $p_1 = 0.1$ MPa and $p_2 = 0.2$ MPa is often used to evaluate the compressibility of soil. For low compressibility soil, $E_s > 15$ MPa; medium compressibility soil, $4\text{ MPa} < E_s \leq 15$ MPa; high compressibility soil, $E_s \leq 4$ MPa.

The Poisson's ratio of soil refers to the ratio of the lateral expansion strain ε_x to the vertical compression strain ε_z when the soil is compressed under the unconfined condition, formula:

$$\mu = -\frac{\varepsilon_x}{\varepsilon_z} \quad (5\text{-}25)$$

When the $e\text{-}\lg p$ curve of soil is redrawn into the $e\text{-}\lg p$ curve of semi-logarithmic compression curve, its posterior segment is close to a straight line (Fig. 5.12b). Its slope C_c is called the compression index of soil, formula:

$$C_c = \frac{e_1 - e_2}{\lg p_2 - \lg p_1} \quad (5\text{-}26)$$

Like the compression coefficient a, the greater the compression index C_c, the higher the compressibility of soil. Different from coefficient a, the C_c does not vary with pressure in the line range, and slope must be very exact, otherwise C_c value error will be great. C_c value of low compressibility soil is generally less than 0.2, and the C_c value greater than 0.4 generally belongs to high compressibility soil. The $e\text{-}\lg p$ curve can be used to analyze the influence of stress history on soil compressibility, which has practical significance for settlement calculation of buildings.

(2) Shear strength of soil

The strength of soil is one of the basic problems of soil mechanical properties. In engineering practice, the strength of soil involves the bearing capacity of foundation.

The stability of the embankment, the slope of the earth dam and the natural soil slope, and the earth pressure and rock pressure acting on the structure when the soil is the environment of the engineering structure, the failure of the soil under the normal stress state, is plastic failure, or shear failure. The shear strength of soil refers to the ultimate capacity of soil to resist shear failure, which is equal to the shear stress on the sliding surface.

The shear strength of soil is usually measured by direct shear test or triaxial shear test. A large number of tests show that under the load of general buildings (0.1~0.6 MPa), the relationship between shear strength and normal compressive stress of soil is approximately linear. The shear strength curve of non-cohesive soil is a straight line passing through the coordinate origin. The equation is:

$$\tau_f = \sigma \tan \varphi \quad (5\text{-}27)$$

The shear strength curve of cohesive soil is an approximate straight line that does not pass through the origin of coordinate and has an approximate straight-line with an intercept C from the ordinate. The equation is:

$$\tau_f = \sigma \tan \varphi + C \quad (5\text{-}28)$$

In the formula: τ_f is the shear strength of soil (MPa); σ is the normal pressure (MPa) on the shear surface; φ is the internal friction angle (°) of soil; C is cohesion of soil (MPa).

The above formula shows that the shear strength of soil is composed of internal friction $\sigma \tan \varphi$ and cohesion C. The internal friction is proportional to the normal pressure of the shear surface, and the ratio is the internal friction coefficient of the soil $\tan \varphi$. The shear strength of non-cohesive soil is determined by the internal friction $\sigma \tan \varphi$, which is proportional to the normal pressure. The internal friction coefficient of soil is mainly determined by the roughness of soil particle surface and the staggered arrangement of soil particles. The rougher the soil particle surface, the more angular and the greater the compactness, the greater the internal friction coefficient of soil. The shear strength of cohesive soil is composed of internal friction and cohesion. The internal friction of soil is mainly composed of water-binder connection or capillary water connection formed by bound water between soil particles.

(3) Dynamic characteristics of soil

The above is the mechanical properties of soil such as compressibility and shear strength under static load, however, under the vibration of shock or machine foundation, a series of physical and mechanical phenomena different from those under static load will occur. In general, the shear strength of soil under dynamic load will decrease, and additional deformation will often occur.

The degree of shear strength decrease and deformation increase of soil under dynamic load depends not only on the type and state of soil, but also on the amplitude, frequency and shock (or vibration) acceleration of dynamic load.

5.5 Engineering Classification of Soil

At present, the widely used engineering classification standards of soil in China mainly includes the classification of *Code for Design of Building Foundation* (GB 50007—2011), *Code for Investigation of Geotechnical Engineering* (GB 50021—2001) (2009 edition) and *Standard for Engineering Classificantion of Soil* (GB/T 50145—2007). The division principles and standard are as follows.

5.5.1 By Accumulation Age

Paleo-deposited soil: Quaternary late Pleistocene Q_3 and its previous accumulated soil, generally in over-consolidation state, high in structural strength;

Ordinary deposited soil: Quaternary Holocene (pre-cultural Q_4) accumulated soil;

Recently deposit soil: the newly accumulated soil layer Q_4, generally under-compacted, and low in structural strength since the cultural period.

5.5.2 By Geological Origin

According to geological origins, soil can be divided into residual soil, diluvial soil, diluvial soil, alluvial soil, glacial soil and aeolian soil.

5.5.3 By Particle Size Distribution or Plasticity Index

Soils in *Code for Investigation of Geotechnical Engineering* (GB 50021—2001) (2009 edition) and *Code for Design of Building Foundation* (GB 50007—2011) are classified into gravel soil, sandy soil, silty soil and cohesive soil according to particle gradation or plasticity index. *Standard For Engineering Classification of Soil* (GB/T 50145—2007) divides soil into mega-grained soil, coarse-grained soil and fine-grained soil.

(1) Gravel soil

Gravel soil refers to soils in which particles larger than 2 mm accounts for more than 50% of the total mass. According to particle size distribution and formation, gravel soil can be divided into boulder, block stone, pebble, gravel, round gravel and breccia (Table 5-10).

(2) Sandy soil

Sandy soil refers to soils in which the mass of particles larger than 2 mm is less than 50% of the total mass, and the mass of particles larger than 0.075 mm is more than 50% of the total mass. According to the particle gradation, it is divided into gravel sand, coarse sand, medium sand, fine sand and silt sand according to Table 5-11.

Classification of gravel soil Table 5-10

Name	Particle shape	Distribution of grain size
Boulder	Circular and sub-circular	Particles larger than 200 mm account for over 50% of total mass
Block stone	Angular	
Pebble	Circular and sub-circular	Particles larger than 20 mm account for over 50% of total mass
Gravel	Angular	
Round gravel	Circular and sub-circular	Particles larger than 2 mm account for over 50% of total mass
Breccia	Angular	

Note: Classification should be based on the particle fraction content from top to bottom to the first match.

Classification of sandy soil Table 5-11

Name	Distribution of grain size
Gravel sand	Particles greater than 2 mm accounts for 25%~50% of total mass
Coarse sand	Particles greater than 0.5 mm accounts for 50% of total mass
Medium sand	Particles greater than 0.25 mm accounts for 50% of total mass
Fine sand	Particles greater than 0.075 mm accounts for 85% of total mass
Silty sand	Particles greater than 0.075 mm accounts for 50% of total mass

(3) Silty soil

Silty soil refers to soils in which the mass of particles larger than 0.075 mm is less than 50% of the total mass, and the plasticity index is less than or equal to 10. According to the particle size distribution, sandy soil can be divided into sandy silt (particle size less than 0.005 mm particle mass is less than 10% of the total mass) and clay silt (particle size less than 0.005 mm particle mass is equal to or more than 10% of the total mass), as shown in Table 5-12.

(4) Cohesive soil

Soil with plasticity index greater than 10. According to the plasticity index, cohesive soil can be divided into silty clay ($10 < I_p \leq 17$) and clay ($I_p > 17$).

5.5.4 By Organic Matter Content

According to the organic matter content can be divided into inorganic soil, organic soil, peaty soil and peat, as shown in Table 5-13.

Classification of silty soil Table 5-12

Name	Distribution of grain size
Sandy silt	Particles less than 0.005 mm account for less than 10% of the total mass
Clay silt	Particles less than 0.005 mm account for 10% or more than 10% of the total mass

Classification of soil by organic matter content　　Table 5-13

Name	Organic content W_u	In-situ identification characteristics	Explanation
Inorganic soil	$W_u < 5\%$	—	—
Organic soil	$5\% \leqslant W_u \leqslant 10\%$	Deep gray, shiny, smelly. After soaking, bubbles appear on the water surface, and the volume shrinks after drying	1. If on-site identification or regional experience is available, organic matter content determination is unnecessary. 2. When $w > w_l$, $1.0 \leqslant e < 1.5$, called silty soil. 3. When $w > w_l$, $e \geqslant 1.5$ is silt
Peaty soil	$10\% < W_u \leqslant 60\%$	Deep gray or black, smelly. Incomplete decomposition of the plant structure can be seen, the volume increases when soaking, there are plant residues floating in water, the volume shrinks after drying	According to regional characteristics and needs, W_u can be divided into: weak peat soil: ($10\% < W_u \leqslant 25\%$) medium peat soil: ($25\% < W_u \leqslant 40\%$) strong peat soil: ($40\% < W_u \leqslant 60\%$)
Peat	$W_u > 60\%$	In addition to the characteristics of peat soil, loose in structure, soil is very light, dark, dry the volume shrinks after drying	Peat soil moisture content may be greater than 100%.

Note: Organic matter content W_u is determined by ignition loss test.

5.6 Engineering Geological Properties of General Soil

5.6.1 Gravel Soil

The gravel soil is mainly composed of rock debris or primary minerals such as quartz and feldspar. The particles are coarse and have a single grain structure. It often has the characteristics of large pores, strong permeability, low compressibility and high shear strength. These properties are related to the content of coarse particles and the property and quantity of fillings in pores. The typical flow sedimentary gravel soil sorting is good, the pore filling is mainly sand of a small amount, so its permeability is strong, compressibility is low, and strength is high.

The sorting of gravel that generated from bedrock weathering and mountain slope accumulation is poor. Soils whose pores are filled with a large number of fine particles such as sand, silt and clay, its permeability is relatively weak, the shear strength is low, and the compressibility is slightly high. In general, gravel soil is a good foundation for general buildings. However, due to its strong permeability and no connection between particles, there are often a series of engineering geological problems such as leakage of dam foundation, channel, reservoir, water gushing of foundation pit and underground tunnel, slope collapse and instability. Gravel is a good choice as filling materials.

5.6.2 Sandy Soil

Sandy soil is mainly composed of primary minerals such as quartz, feldspar and mica. Generally, it is not connected and has a single grain structure, with strong permeability, low compressibility and high strength. These properties are related to sand size and density. The general properties of coarse and medium sand are good, which can be used as a good foundation for buildings. But it may cause engineering geological problems such as water gushing or leakage. The engineering geological properties of fine sand and silty sand are relatively poor, and such sands prone to liquefaction when subjected to vibration, and cause flow sand when excavation, which will endanger the safety of buildings. Fine sand and silty sand are generally not suitable as concrete aggregate.

5.6.3 Cohesive Soil

The clay content in cohesive soil is high, and it often contains hydrophilic clay minerals with water-binder connection and agglomeration structure, and the pores are small and in large amount. Cohesive soil with different moisture contents often shows different consistency states. Its compression rate is small, and compression amount is large. The shear strength mainly depends on the cohesion, and the internal friction angle is small. The engineering properties of cohesive soil mainly depend on its connection and compactness, which is related to clay content, consistency and pore ratio. The plasticity, swelling, compressibility, permeability and shear strength of cohesive soil can change greatly with the increase of clay content. Consistency can have the greatest impact on cohesive soil properties, and the cohesive soil near the flow state and soft plastic state has higher compressibility and lower shear strength. The cohesive soil in solid or hard plastic state has lower compressibility and higher shear strength.

Quiz

1. According to the causes of soil classification, what categories can soil be divided into? What are the basic characteristics of various soils?

2. What is the structure of soil? What are the characteristics of different soil structures?

3. How to understand the three-phase composition of soil?

4. What are the forms of water in soil? How does it affect the engineering properties of soil?

5. What are the physical state indexes of clay-free soil and clay soil?

6. What are the mechanical indexes of soil?

7. Why are the tight state of cohesive soil and the plasticity index and liquid index of cohesive soil comprehensive indicators reflecting their respective engineering properties?

8. How to classify engineering soil and rock in China?

9. What are the general engineering geological characteristics of gravel soil, sand soil and clay soil? What kind of engineering geological properties it is?

Chapter 6
Groundwater and its Engineering Effects

6.1 Basic Conceptions

Groundwater is a common topic in construction projects. Groundwater refers to the water that exists and migrates in the rock and soil voids below the ground.

Therefore, in order to consider the reaction consequences of engineering process on the ground to groundwater, it is necessary to consider the role of groundwater in those processes and reactions. Here, four terms used in hydrogeology need to be explained.

① Water table: the level of standing water in the ground;

② Aquifer: strata in which water has a great significance to engineering practice (aquifers are described as unconfined if their upper surface is ground, and confined if they are covered-or "confined" -by less permeable material);

③ Aquitard: strata through which water can penetrate at a certain rate;

④ Hydro-geological boundary: a surface (bedding, fault, etc.) that separates the strata with different hydro-geological characteristics.

In a whole, groundwater is an extremely important factor in engineering geological analysis, evaluation and geological disaster prevention. The study of groundwater is mainly about the hydro-geological conditions, such as the type of groundwater, burial, recharge, runoff and discharge conditions, the permeability of rock and soil, and the influence of groundwater on projects.

6.2 Physical Properties and Chemical Composition

6.2.1 Physical Properties of Groundwater

The physical properties of groundwater include temperature, color, transparency, taste, smell, specific gravity, electrical conductivity and radioactivity, etc.

(1) Temperature of groundwater

The temperature of groundwater is controlled by climatic and geological conditions. Because of the different environment in which groundwater is formed, its temperature varies greatly, ranging from 0~100℃, and above 100℃ in some areas. Groundwater is divided into the following categories according to temperature: Super cooled water (<0℃); Extreme cool water (0~4℃); Cool water (4~20℃); Warm water (20~37℃); Hot water (37~42℃); Extreme hot water (42~100℃); Super-hot water (>100℃). The depth of groundwater circulation can also be roughly calculated according to the temperature of the water. The formula is as follows:

$$H = q(T-t)+h \qquad (6-1)$$

H—the depth of groundwater circulation, m;
q—geothermic depth, m/℃;
T—water temperature, ℃;
t—annual average temperature, ℃;
h—the depth of the annual normal temperature zone, m.

(2) Color

The color of groundwater is determined by its chemical composition and suspended matter. For example, water with Ca^{2+} and Mg^{2+} is bluish; water with Fe^{2+} is grayish blue; water with Fe^{3+} is brownish yellow; and water with organic humus is yellow. Water containing suspended matter color is determined by the suspended matter.

(3) Transparency

Most groundwater is transparent. When the water contains minerals, mechanical mixtures, organic matter and colloids, the transparency of groundwater changes. According to transparency, groundwater can be divided into the following categories: ① transparent; ② slightly cloudy; ③ turbid; ④ extremely turbid.

(4) Taste

The taste of groundwater mainly depends on the chemical composition. Water containing NaCl has a salty taste; water containing $CaCO_3$ is refreshing; water containing $Ca(OH)_2$ and $Mg(HCO)_2$ has a sweet taste, commonly known as sweet water. Groundwater has a bitter taste when $MgCl_2$ and $MgSO_4$ are present.

(5) Smell

When groundwater contains gas or organic matter, it has a certain smell. For example, when it contains humus, it has a "swamp" smell; when it contains hydrogen sulfide (H_2S), it has a smell of rotten eggs.

(6) Specific gravity (relative density)

It is decided by the content of dissolved salts. The more salts, the greater specific gravity. The relative density of pure water is 1.0, or even up to 1.2~1.3.

(7) Electrical conductivity

It is decided by the quantity and properties of electrolytes. The more ions, the higher ionic valence, the stronger electrical conductivity.

(8) Radioactivity

It depends on the content of radioactive material. In general, underground water are all radioactive.

6.2.2 Chemical Composition of Groundwater

(1) Chemical composition

In the quality analysis of groundwater, in addition to its physical properties, the chemical composition of groundwater should also be determined. Common chemical components in groundwater are listed in Table 6-1.

The so-called chemical composition of groundwater refers to the dissolved salts and gases in groundwater. The solubility of several salts (at 18℃) is listed in Table 6-2.

It should be noted that the value of solubility not only varies with temperature and

Common chemical components in groundwater   Table 6-1

Elements and genetic components	Enrichment conditions of elements in water	Chemical elements
Gas	Igneous activity, biological origin, air origin, chemical origin, radioactive origin	O_2; N_2, CO_2, CH_4, H_2S, HCl, HF, H_2S, SO_3; Inert gas Ar, Ne, He, Kr, Xe, Rn, Th, O_3, N_2O; SO_2, SO_3, Cl;
Major ions, molecules, micronutrient (content<10^{-3}%)	All kinds of causes	Cl^-, SO^{4-}, HCO^{3-}, CO^{3-}, NO^{3-}, Na^+, K^+, Ca^{2+}, Mg^{2+}, H^+, NH^{4+}, H_3SiO_4, Fe^{2+}, Fe^{3+} and organic matter, etc.
	Various causes of formation: enrichment of metallic elements in the oxidation zones of pyrite, copper ore and other ore deposits with decreasing pH value, enrichment of ammonium iodide and enrichment of metallic elements in areas where oil, gas fields and other organic matter accumulate, enrichment of Li, F, Br, silicic acid and other micronutrient occurs in groundwater in crystalline rock areas	Li, Be, F, Ti, V, Cr, Mn, CO, Ni, Cu, Zn, Ge, As, Se, Br, Rb, Sr, Zr, Nb, MO, Ag, Cd, Sn, Sb, I, Ba, W, Au, Hg, Pb, Bi, Tb, U, Ra

Continued

Elements and genetic components	Enrichment conditions of elements in water	Chemical elements
Colloid	Positive colloid	$Fe(OH)_3, Al(OH)_3, Cd(OH)_2, Cr(OH)_3, Ti(OH)_4, Zr(OH)_4, Ce(OH)_3$
	Negative colloid	Viscous colloid, Humus, $SiO_2, MnO_2, SnO_2, V_2O_3, Sb_2S_3, PbS, As_2S_3$
Organic Matter (bacteria)	Metabolic products, life and death decomposition of humic acid and fulminate bacteria	organic polymer compounds, Humic acid (fulminic acid C:44%; H:53%; O:40%; N:15%), Algae medium, bacterial, Humus, rafaelite, phenol, fatty acid, naphthenic acid

Solubility of common salts in nature   Table 6-2

Salts	Solubility	
	g / L	Molal concentration
NaCl	328.6	5.42
Na_2SO_4	168.3	1.15
Na_2CO_3	193.9	1.8
KCl	329.5	3.9
K_2SO_4	111.1	0.62
K_2SO_3	1080.0	5.9
$MgCl_2$	558.1	5.1
$MgSO_4$	354.3	2.8
$CaCl_2$	731.9	5.4
$CaSO_4$	2.0	0.015

pressure, but also with the presence of other salts or some gases.

(2) Chemical property

1) Total salinity: it is the total content of all ions in groundwater. When the groundwater is heated to evaporate (105~110 ℃), the number of solid residuals is called the total salinity.

2) pH of groundwater: the pH of groundwater depends mainly on the concentration of hydrogen ions in water. Its pH value is expressed by the negative logarithm of H^+ concentration. pH=-Lg [H^+]. According to the pH value of groundwater, the water is divided into five classes (Table 6-3).

3) Hardness of groundwater (Table 6-4)

Hardness is usually divided into total hardness, temporary hardness and permanent hardness. The total content of Ca^{2+} and Mg^{2+} in water is called total hardness. After water boiling, some of the bicarbonate of Ca^{2+} and Mg^{2+} in the water lost CO_2 and formed carbonate precipitation, which can reduce the contents of Ca^{2+} and Mg^{2+} in the water. The total contents of Ca^{2+} and Mg^{2+} after decreasing due to boiling are called

Acid-base properties: value of pH Table 6-3

Class	Strong acid water	Weak acid water	Neutral water	Weak alkaline water	Strong alkaline water
pH	< 5.0	5.0~7	7	7~9	>9

Water hardness Table 6-4

Type of water	Water hardness		
	Germanic degree	meq/L	mol/L
Extreme soft water	<4.2	<1.5	$<7.5\times10^{-4}$
Soft water	4.2~8.4	1.5~3.0	7.5×10^{4}~1.5×10^{8}
Slight hard water	8.4~16.8	3.0~6.0	1.5×10^{-3}~3×10^{-3}
Hard water	16.8~25.2	6.0~9.0	3×10^{-3}~4.5×10^{-3}
Extreme hard water	>25.2	>9.0	$>4.5\times10^{-3}$

temporary hardness, also known as carbonate hardness. The difference between the total hardness and the temporary hardness is called permanent hardness, which is the content of Ca^{2+} and Mg^{2+} that don't experience carbonate precipitation during boiling. Generally, when the hardness of the water is too low, and it will erode the concrete.

Germanic degrees are commonly used to express hardness of groundwater. 1 degree = 10 mg (CaO)/L or 7.2 mg (MgO)/L.

6.3 Classification of Groundwater

6.3.1 Aquifer and Aquiclude

An aquifer (Fig. 6-1) is a rock or soil layer that not only contains water but from which water can be abstracted readily in significant quantities. The ability of an aquifer to transmit water is governed by its permeability (Fig. 6-2).

6.3.2 Types and Characteristics of Groundwater

The methods of groundwater classification are various, and there are mainly two: ① According to the burial conditions, groundwater is classified into three types:

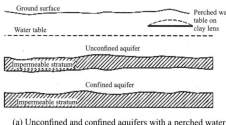

(a) Unconfined and confined aquifers with a perched water table in the vadose zone

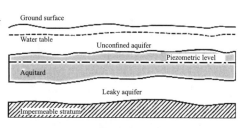

(b) A leaky aquiferl

Fig. 6-1 Aquifer

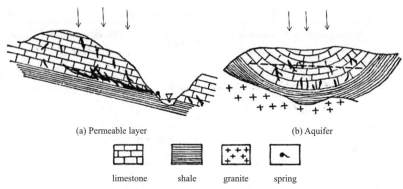

Fig. 6-2 Conditions for the formation of aquifers

aeration zone water, phreatic water and confined water. ② According to the pore nature of the aquifer, the groundwater is divided into pore water, fracture water and karst water.

(1) Aeration zone water

Aeration zone water: the temporary stored in soil above phreatic water table, including soil water, marsh water and stagnant water.

1) Soil water

Soil water (Fig. 6-3) is located in the uppermost part of the Earth's crust and mainly exists in the soil layer in the form of pellicular water, hydroscopic water and capillary water.

Fig. 6-3 Soil water

2) Marsh water

Marsh phenomenon will occur when atmospheric precipitation far exceeds evaporation, and the water permeability of the soil is poor, or when the phreatic level rises to a level to connect with soil water. The water in the swamp is called marsh water, which is stored in the marshland, sometimes in grass root layers and peat layers (Fig. 6-4).

3) Perched water

Vadose zone is also called unsaturated zone, which contains bound water, capillary water and gaseous water. The vadose water is affected by the adsorption force on the surface of the particles, the capillary tension of the pores and the gravity. The gravity water distributed on the surface of the partially impermeable or impermeable layer in the aeration zone is called the perched water (Fig. 6-5).

Fig. 6-4 Marsh water

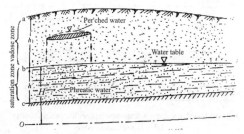

Fig. 6-5 Perched water

Formation conditions of perched water: ① There are partial or impermeable aquifers in the permeable layer; ② The attitude of the aquifer should be level or close to level; ③ The aquifer has a certain range.

Features of perched water: ① Controlled by climate, water volume is small, unstable, and seasonal changes are obvious; ② The recharge source is atmospheric precipitation and surface water infiltration, and the recharge area is consistent with the distribution area; ③ The water table is shallow, easy to evaporate, pollution, and water quality is poor.

Engineering significance of perched water: ① It is not a good water supply choice, but it can be used as a small water supply source in water-deficient areas, such as the loess plateau; ② The presence of unsaturated water can weaken the strength of the foundation soil; ③ In cold areas, it can easily cause frost heaves and mud boiling. ④ The great changes of groundwater table often bring difficulties to engineering design and construction.

Treatment methods of perched water: Drain water in the vadose zone, and pierce the bottom of the water barrier and discharge to the lower aquifer.

(2) Phreatic water

Phreatic water is the groundwater above the first stable aquiclude below the surface of the earth (Fig. 6-6).

1) Features

Phreatic water has the following characteristics according to the burial conditions:

① Phreatic water is an unpressurized water with a free surface, which is connected with the ground surface through the vadose zone, while vadose zone water has no stable water-resisting layer.

② The burial depth and the thickness of the aquifer of phreatic water are affected by climate, topography and geological conditions, and vary greatly.

③ The phreatic water flows from high to low under the action of gravity, and the velocity of flow depends on the permeability and hydraulic gradient of the strata.

④ In general, the distribution of phreatic water is consistent with the recharge zone, precipitation and surface water infiltration are the main recharge sources of phreatic water. In addition, the condensed water is also one of the important sources in the arid and semi-arid areas with great day-night temperature difference. The drainage of phreatic water is mainly by evaporation and horizontal drainage to adjacent aquifers or to surface water bodies.

⑤ Phreatic water takes an active part in the water cycle. Its resources are easy to be replenished and restored. The dynamic of phreatic water is greatly influenced by climate. Phreatic aquifer is generally limited in its thickness, and its resources usually lack multi-year accommodation.

⑥ The quality of phreatic water is mainly determined by climatic, topographic and lithologic conditions.

2) Supply conditions for phreatic water

The main sources of replenishment (Fig. 6-7) for phreatic water are atmospheric precipitation, rain, melted snow, dew and so on. The amount of recharge phreatic water depends on the characteristics of precipitation, permeability, opening degree and plant cover of surface rock.

In arid regions, rivers usually provide replenishment to phreatic water. Sometimes they may complement each other. When the water table of the river is lower, phreatic water will replenish the river (Fig. 6-8a). When the water table of the river is higher, the river water will recharge phreatic water (Fig. 6-8b). There are also cases where one side of the

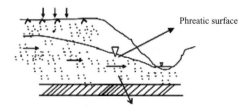

Fig. 6-6 Phreatic water

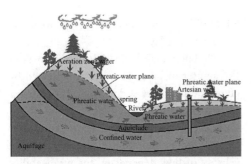

(a) Atmospheric precipitation replenishment

(b) Surface water replenishment

Fig. 6-7 Water replenishment

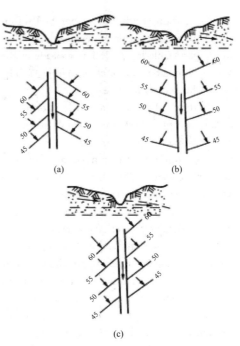

Fig. 6-8 Relations between phreatic water and river water

river is recharged by phreatic water, while the other side, river is recharging phreatic water (Fig. 6-8c). The recharge relationship between phreatic water and surface water can be determined by the groundwater table.

3) The engineering evaluation of phreatic water

Phreatic water has an effect on the stability of buildings because of shallow buried water, and it is one of the main factors of construction difficulties, such as quicksand and foundation pit surge. In engineering, the usual treatment schemes are as follows: the building foundation is best selected in the zone with deep phreatic water table; drainage and lowering of water level can be adopted when the construction of the foundation is affected, isolation or freezing method construction and other measures to deal with .

Variation in the lithology or thickness of an aquifer: when the grade of the topographic does not change much, the permeability of the aquifer become better or the thickness of the aquifer become greater if the isohydrous line becomes thinner (Fig. 6-9). On the contrary, the permeability of the aquifer has become worse or its thickness has become smaller.

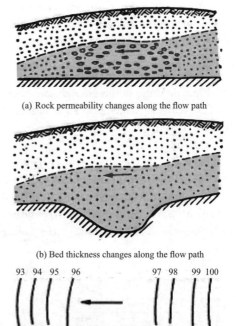

Fig. 6-9 Variation in the lithology or thickness of an aquifer

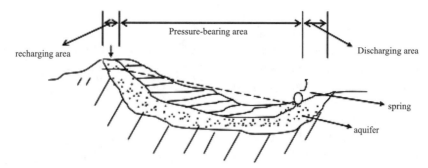

Fig. 6-10 Confined water

4) Basic types of phreatic water

Phreatic water is widely distributed in Quaternary loose sediments, and also in fractured rocks in bedrock areas. According to the different burial conditions of phreatic water, it can be divided into six main types: river valley phreatic water, phreatic water in glacial sediments, phreatic water in the piedmont plain, phreatic water in grasslands, phreatic water in deserts and semi-deserts, phreatic water in coastal sands and bedrock.

(3) Confined water

1) Features

① The important feature of confined water is that there is no free water surface, it has confined property and bears certain hydrostatic pressure.

② The distribution of replenishment area, pressure bearing area and drainage area is often obvious, and the replenishment area is relatively far away from the drainage area. The recharge area is the exposed surface part of the confined aquifer, which can accept the recharge of precipitation, surface water and upper perched water, and has the characteristics of hydrostatic pressure (Fig. 6-10).

③ Limited area of confined water is connected with the outside, its participation of the water cycle is not as active as phreatic water, water alternates slowly, the average residence time is long (old or long).It is not easy to be replenished, or restored, thus little change in water quality, temperature, quantity, chemistry and so on. And it enjoys dynamic stability, and has multi-year regulation.

④ The quality of confined water depends on the burial conditions and the degree of its connection with the outside. It can be fresh water or brine with a high salt content. Usually the water quality is good, the water quantity is stable, and it is not easy to be polluted, but difficult to repair after pollution. It is a good source of water supply.

⑤ The replenishment method of confined water is much more complicated than phreatic water. The main replenishment sources include atmospheric precipitation infiltration recharge, surface water infiltration recharge, and adjacent aquifer recharge. The discharge of confined water is mainly in the form of springs, and replenishment of surface water bodies and adjacent aquifers.

2) Burial conditions of confined water

The formation of confined water is mainly determined by geological structures. Therefore, the burial conditions of confined water are determined by different geological structural forms, which is the main difference between confined water and phreatic water.

① Syncline structure: syncline structure is often the most favorable place for the existence of confined water, which forms a complete confined basin.

The relationship between topography and structural morphology can also reflect the interrelationship of recharge, pressure, and drainage of confined water at different depths. When the topography is consistent with the structural form (called normal topography), the water table of the lower aquifer is higher than that of the upper aquifer, which makes it possible for the lower confined water to recharge the upper aquifer. At this point, an confined basin is formed if the artesian pressure head is sufficient to allow groundwater to eject to the surface (Fig. 6-11a). On the contrary, if the topography is inconsistent with the structural form (called negative topography), the confined water in the upper aquifer may recharge the lower aquifer (Fig. 6-11b).

② Monoclinic structure: the strata with monoclinic structure can also form monoclinic bearing-pressure slope under proper conditions.

③ Piedmont phase transition: alluvial deposits in the piedmont zone often form a pinch-out aquifer and a piedmont confined slope. Part of the alluvial deposits in the front of the mountain are mainly pebbles and gravel, forming a phreatic water zone, which is a recharge area for the artesian slope. Differences between phreatic water and confined water are shown in Table 6-5.

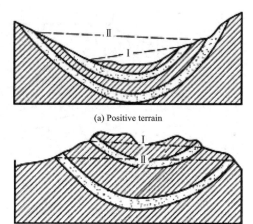

(a) Positive terrain

(b) Negative terrain

Fig. 6-11 Types of confined basins

(4) Fracture water

The groundwater buried and transported in various fractures of rock mass, e.g., joints and cracks, is called fracture water (Fig. 6-12).

1) Features

Water-bearing capacity and water-rich capacity are very uneven, the water difference is great. The main influencing factors are the

Differences between phreatic water and confined water   Table 6-5

Items	Phreatic water	Confined water
Storage conditions	Stored above the first aquifuge	Stored between two aquifuges, and bearing some static pressure
Recharging	Precipitation and river	Precipitation, river and phreatic water
Discharging	Form springs when exposed to the ground, or replenish surface water or evaporate	Replenish phreatic water or surface water, form springs when exposed to the ground
Features	① With free water surface; ② Flow along incline direction under gravity due to the ground slope; ③ Distribution area is coincide with recharging area ④ Storage depth is shallow, table and flow quantity is unstable; ⑤ Affected by weather, easy to be polluted	① Confined by aquifuges, bearing static pressure which controls the movement of confined water; ② Distribution area, recharging area and discharging area are not in the same region; ③ Storage in relative deep strata, table and flow quantity are stable; ④ Not affected by weather, uneasy to be polluted

Fig. 6-12 Fracture water

cause of formation of fractures, the degree of development of fractures, filling situation, connection situation, source and condition of groundwater supply.

2) Classification

① According to burial conditions: there are three types, which are fracture perched water, fracture phreatic water, and fracture confined water.

② According to the attitude of the igneous rock: there are two types, which are bedded fracture water (such as layered aquifer, groundwater with uniform water level in weathered fissures); vein fracture water (fault fractures zone aquifer).

③ According to different causes: there are three types, which are weathered fracture water, rock-formation fracture water and structure fracture water.

(5) Karst water

In the soluble strata composed of limestone, dolomite, gypsum, and rock salt, the groundwater flow continuously undergoes filtration and dissolution occurs when groundwater flows through rock cracks, pores, etc., and wide crevices and karst caves of varying volumes will be created.

Engineering problems caused by karst water are as following:

① In most coal seams in North China, the high pressure head of the underlying limestone water brings huge mining risks.

② There is severe water shortages in karst mountainous areas in Yunnan, Guangxi and Guizhou provinces in China suffer from, hindering development and causing poverty.

③ The reservoir water seepage in the karst area is difficult to deal with.

④ Underground construction and tunnel construction are difficult to be carried out.

(6) Spring

1) General concept

The natural outcrop of groundwater gushing out of the surface is called spring.

2) Genetic classification

According to the supply source, springs can be divided into three types:

① Vadose spring: Such springs are located in the aeration zone and supplied by the perched water. Therefore, it is characterized by great variations in water inflow and is largely dependent on local climatic conditions. It can dry up in some seasons. The chemical composition and temperature of this spring water are also unstable.

② Depression spring: Such springs are supplied by phreatic water, and their dynamics are more stable and regular than vadose springs. There are seasonal changes in water volume, water temperature, and chemical composition.

③ Ascending spring: The supply source of such spring water is confined water. Its appearance is characterized by spring water upwelling, which is dynamic and stable, and the amount of water inflow varies little throughout the year.

According to the reasons of spring exposure, springs can be divided into four types: erosion spring, contact spring, overflow spring and fault spring.

When the river erosion net is cut down to reach the aquifer, groundwater flows out of the surface and appears along the river on both banks. This kind of spring formed by erosion is called erosion spring. Erosion depression spring are formed when the aquifer is phreatic (Fig. 6-13a) and erosion ascending springs are formed when the aquifer is confined (Fig. 6-13b). Contact springs are formed because the permeable layer is locally connected with the water barrier, or formed by groundwater rising to the surface along the feature in the contact zone between the intrusive body and the surrounding rock (Fig. 6-13c, d). Besides, there are fault springs in

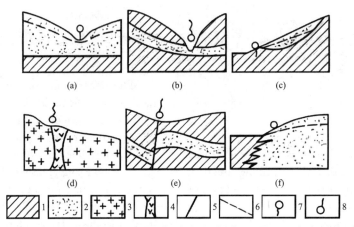

1—Aquifuge; 2—Permeable layer; 3—Hard rock; 4—Rock vein; 5—Water-conducting fault;
6—Underground water table; 7—Depression spring; 8—Ascending spring

Fig. 6-13 Schematic diagram of spring formation conditions

linear distribution caused by faulting.

(7) Groundwater in multi-year freezing zone

1) The concept of multi-year freezing

Permafrost layers are mainly distributed in areas where the annual average temperature is below zero degrees Celsius.

2) The basic type of multi-year groundwater in the freezing zone

The former USSR scholar H. N. Tolzzikhin divided the groundwater in the multi-year freezing zone into three categories:

① Water above frozen layer: It is mainly buried in the thawing layer that freezes in winter and melts in summer. From the topography, it is a depression area and a gentle watershed area. The thickness of the thawing layer is related to the topography and climate of the area.

② Water between frozen layers: This refers to the groundwater between the upper boundary and the lower boundary of the multi-year freezing zone.

③ Water in the base layer of frozen soil: This refers to the water below multi-year frozen layers. Its main feature is that it always maintains a liquid phase.

6.4 Flowing of Groundwater

6.4.1 Basic Laws of Groundwater Movement

In saturated gravel and fractured rocks, groundwater is affected by gravity (differential head) and flows from places with high water table to places with low water table. The movement of gravitational water in porous medium is penetration.

The movement law of groundwater in natural condition depends on various conditions of the aquifer boundaries and that of the aquifer interspace. According to present researches, the movement of groundwater in saturated rock and soil has three kinds: laminar flow, turbulent flow and mixed flow. It must be pointed out that there is still a lot of controversy about whether there is mixed flow infiltration under natural conditions.

(1) Law of linear penetration

Before attempting any mathematical treatment of groundwater flow, simplifying assumptions must be made. ① The materials are isotropic and homogeneous. ② There is no capillary water effect. The flow state is stable. ③ Since rocks and soils, anisotropic and heterogeneous, may be subject to capillary action and flow through them is characteristically unsteady, any mathematical assessment of flow must be treated with caution.

H·Darcy (1856) enunciated the basic law concerned with flow, which states that the rate of flow, v, per unit area is proportional to the gradient of the potential head, I, measured in the direction of flow, and K is the coefficient of permeability:

$$v = KI \quad (6-2)$$

The acquisition of various data in the above formula is related to the experimental device.

$$v = \frac{Q}{W} \quad (6-3)$$

$$I = \frac{H_1 - H_2}{L} \quad (6-4)$$

v — permeability rate, m/s;
K — permeability coefficient, m/s;
I — hydraulic gradient (indicate the water head loss in unit traveling distance);
Q — rate of flow, m³/s;
W — cross-sectional area (Darcy's cross-sectional area), m²;
H_1, H_2 — the head of the upper and lower section of the underground water, m;
L — traveling distance (the length of the sand column in the instrument), m.

According to the definition, the seepage flow velocity v is quantitatively equivalent to the seepage flow through a unit section within a unit time.

If $I = 1$, then

$$v = K \quad (6-5)$$

This means that the coefficient of permeability is quantitatively equal to the value of the seepage velocity per unit area at a hydraulic gradient $I = 1$. Therefore, the permeability coefficient has the dimension of velocity, which is usually expressed as m/d, m/h and cm/s. According to the K value, the aquifer can be divided into four types: strong permeable layer (>10 m/d), permeable layer (10~1 m/d), weak permeable layer (1~0.001 m/d) and impervious layer (<0.001 m/d). As shown in Table 6-6.

Permeability coefficient of common soil
Table 6-6

Soil name	Permeability coefficient K (m/d)
pure gravel	200
pure pebble	100~200
gravelly coarse sand	50~100
coarse sand	25~75
medium sand	10~25
fine sand	23~10
fine clay sand	1~2

It should be noted that the above permeability velocity v does not represent the real velocity of water movement in the pores or cracks of rocks, but is a quoted value, equivalent to the velocity generated under the condition that there is no granular framework in the entire section. In addition, under the same natural factors, the permeation velocity is smaller than the actual velocity u of water flowing in the rock pores. The actual speed is related to the area of pores and cracks in the entire rock section, which it is related to the porosity n. Let W_0 be the actual cross section, the actual speed can be expressed by the following formula.

$$u = Q/W_0 = Q/(nW) = v/n \quad (6-6)$$

$$n = W_0/W < 1 \quad (6-7)$$

Permeability rate is always lower than actual penetration rate.

Darcy's law is a basic law of permeation followed by the movement of groundwater in porous media—the law of linear permeation. It is suitable for laminar movement. Darcy's law can be applied in most of the hydrogeological problems in nature. In general concept, clay is a water barrier. However, when the hydraulic gradient of the seepage field reaches a certain value, the clay will also show a certain degree of water permeability (Fig. 6-14). At this time, Darcy's law of permeability can be used as a piecewise function.

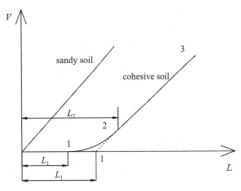

Fig. 6-14 Relationship between seepage velocity of different soil samples and hydraulic gradient

$$v = 0 \quad (I \leq I_1) \quad (6\text{-}8)$$

$$v = Ke^{(I-A)/B} \quad (I_1 < I < I_2) \quad (6\text{-}9)$$

$$v = K(I - I_1') \quad (I \geq I_2) \quad (6\text{-}10)$$

I_1— non-linear initial head gradient;
I_2— linear initial head gradient;
I_1'— the intercept of the straight line of the I axis;
A, B —the undetermined value determined by the penetration test.

(2) Law of non-linear permeability

1) Turbulent flow

The movement of groundwater in rocks strongly destroyed by huge fractures systems and karst caves often does not have a laminar flow. The movement of water in these huge cracks and holes is similar to the nature of water movement in large pipes and channels: the speed of water is very high, and the small streams interfere with and mix with each other and generate vortices. This kind of motion is called turbulent movement in hydraulics.

For groundwater moving in turbulent nature, its permeability rate can be calculated as follows.

$$v = K_0 I^{1/2} \quad (6\text{-}11)$$

K_0—the turbulent permeability coefficient of the rock strata.

Under turbulent flow conditions, the penetration velocity is proportional to the 1/2 power of the hydraulic gradient.

2) Mixed-flow

The movement of mixed flow can be expressed by Scheidegger relation.

$$v = K_H I^{1/m} \quad (6\text{-}12)$$

K_H—The mixed flow permeability coefficient of the rock strata;

m—A certain value between 1 and 2, determined by experiment.

When $m=1$, the above formula takes the form of the linear permeability law. When $m=2$, the above equation becomes the turbulence equation.

To sum up, in fractures or karst rock strata, large fractures, large caves, small fractures and small channels are often staggered. In large caves, groundwater moves in turbulent flow, while in small fractures, the movement of water maintains the nature of laminar flow. When the water flows from a small fracture to a large cave, the form of movement changes. One part of the stream is in the form of turbulent flow, while the other part of the stream still maintains laminar flow. Only under this condition can the movement of groundwater transition from laminar flow to turbulent flow.

However, some data show that as a stable movement law of groundwater in permeable strata, mixed flow does not exist in natural conditions. Therefore, it cannot be compared with the linear permeability law and non-linear permeability law.

6.4.2 Dewatering Problems in Foundation Pit Construction

When the foundation pit is excavated, if the groundwater and surface water flowing into the pit are not removed in time, water will worsen construction conditions, cause soil wall collapse, and reduce the bearing capacity of the foundation. Construction drainage can be divided into two types: catchment well for reducing water table and the artificial lowering of the groundwater table.

Fig. 6-15 shows artificially lower the groundwater table.

Fig. 6-16 shows dewatering design for foundation pit construction.

In addition to the plane layout and elevation layout of the well point system, the dewatering design of foundation pit construction should also calculate the water inflow, determine the number of well pipes and well spacing, and select pumping equipment.

According to the degree of exposure of the aquifer of the vertical water collection structure, it is divided into two types: ① Complete well: a well penetrates the thickness of the aquifer and the bottom of the well places on the aquifer. ② Incomplete wells: wells that do not completely penetrate the entire aquifer or penetrate the entire aquifer, but only part of the thickness of the aquifer is filled with water.

According to the presence or absence of water pressure, it is divided into two types: ① Phreatic well: a borehole or well that exposes the phreatic layer. ② Confined well: a borehole or well that exposes a pressurized water layer.

Fig. 6-17 shows division of well points.

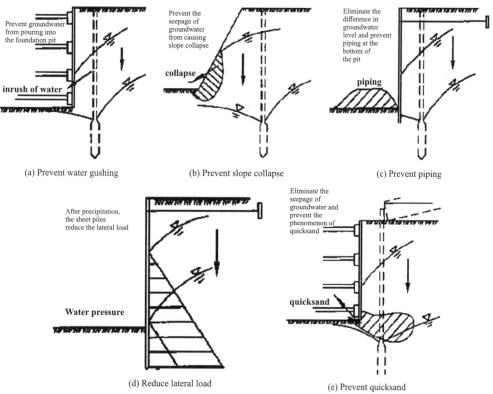

Fig. 6-15 Artificially lower the groundwater level

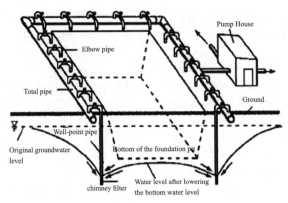

Fig. 6-16 Dewatering design for foundation pit construction

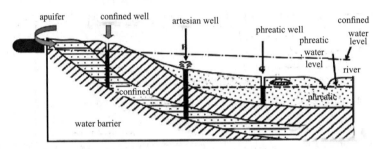

Fig. 6-17 Division of well points

6.5 Impacts of Groundwater on Engineering Construction

6.5.1 Land Subsidence and Collapse

When deep foundation construction is carried out in loose sedimentary layers (such as soft soil along the coast of China), it is often necessary to artificially lower the groundwater table. If the precipitation is not appropriate, the surrounding foundation soil will produce consolidation settlement, which may cause uneven settlement of adjacent buildings or underground pipelines, or even soil particles lost or even hollowed out under the foundation of the building. This will lead to cracking of the building and endangering building safety. (Fig. 6-18)

6.5.2 Seepage Failure of Groundwater

Seepage failure or seepage deformation refers to movement of the whole block or its particles, or the change of its particle composition under the action of seepage. Seepage failure is common in loose soil. Seepage failure may cause cavities in the rock and soil, ground collapse, or spring and sand gushing at the water outlet, thereby affecting the stability of the site, foundation and surrounding rock of the building. A lot of research and practice show that seepage failure mainly includes two basic forms: quicksand and piping.

Fig. 6-18 Land subsidence

(1) Quicksand

Quicksand refers to simultaneous movement of particles or groups of particles within a range of cohesive or incohesive soil under the action of bottom-up seepage (Fig. 6-19) when the seepage force is greater than the gravity of the soil or the hydraulic gradient of the groundwater is greater than the critical hydraulic gradient. Flowing soil occurs at the place of seepage escaping but not inside the soil, such as around the bottom of the deep foundation pit and at the toe of the retaining wall. Quicksand can cause a large amount of soil to flow during the construction of the project, causing the surface to collapse or the foundation of the building to be destroyed. Preventive measures must be taken to prevent such seepage failure.

In areas where quicksand may occur, the above soil should be used as a natural foundation, and pile foundations can also be used to penetrate the quicksand layer. In short, try avoid excavation. If excavation is necessary, the following methods can be used to treat the quicksand:

① Artificially lower the groundwater level: lower the groundwater level below the strata where quicksand may occur, and then excavate;

② Setting up sheet piles: drive sheet piles into the soil, on the one hand, it can strengthen the pit wall, and at the same time increase the seepage distance of groundwater to reduce the hydraulic gradient (Fig. 6-20);

③ Freezing method: freeze groundwater and then excavate;

④ Underwater excavation: use machinery to excavate underwater in the foundation pit (or caisson) to avoid the difference in water head that can cause quicksand. In order to increase the stability of the sand, inject water into the foundation pit and excavated at the same time (Fig. 6-20c).

In addition, chemical reinforcement methods, explosion methods, and weighting methods can be used to deal with quicksand. In the process of excavation of the foundation pit, when quicksand appears in a local area, throw big rocks in the pit immediately to prevent the movement of the quicksand.

Fig. 6-21 shows the damage of quicksand in foundation pits.

(2) Piping

Under a certain hydraulic slope, seepage will cause relatively large hydrodynamic pressure to wash away some of the fine particles in the dissolved soil to form latent corrosion, which can occur under natural conditions and in engineering activities.

The corrosion can be divided into mechanical corrosion and chemical corrosion.

Fig. 6-19 Quicksand

(a) Dry season construction method (b) Setting up sheet piles

(c) Underwater excavation (d) Deep foundation pits use well points

Fig. 6-20 Quicksand treatments

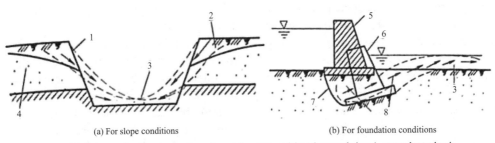

(a) For slope conditions (b) For foundation conditions

1—original slope surface; 2—slope surface after quicksand; 3—quicksand accumulation; 4—groundwater level;
5—building original position; 6—building position after quicksand; 7—sliding surface; 8—quicksand occurrence area

Fig. 6-21 Quicksand damage

Mechanical erosion means that the soil particles are washed away under the action of the hydrodynamic pressure of the groundwater, and the fine particles are washed away, while the larger particles remain in place. The continuous development of this effect will destroy the structure of the soil and form caves. The same phenomenon caused by human engineering activities is also called piping (Fig. 6-22). Chemical erosion refers to the effect of groundwater dissolving the soluble salt in the soil, destroying the binding force between soil particles and the structure of the soil, and the soil particles are taken away by the water to form caves. Chemical erosion is different from karst. The mechanical erosion of seepage is the main one, while chemical dissolution is only subordinate. The

Fig. 6-22 Piping

latter creates conditions for the former. These two effects generally happen at the same time. When groundwater erosion occurs in the foundation soil layer, it will destroy the strength of the foundation soil, form voids, and cause surface collapse, which affects the stability of the construction project. In our country's loess and karst areas, cavitation is common, which should be paid attention to when buildings construction.

The conditions for the generation of piping are as follows.

First, there is a suitable composition of rock and soil particles. Generally speaking, the larger the unevenness coefficient of the rock and soil layer, the easier it is to generate piping. When C_u is greater than 10, piping is likely to occur.

Here, C_u is an index that reflects the uniformity of the particles that make up the soil.

$$C_u = \frac{d_{60}}{d_{10}} \qquad (6\text{-}13)$$

Limited particle size (d_{60}): soil particles smaller than a certain size account for 60% of the total mass of the soil.

Effective particle size (d_{10}): soil particles smaller than a certain size account for 10% of the total mass of soil.

Secondly, there are sufficient hydrodynamic conditions. When the hydraulic gradient of the local seepage flow is greater than the critical hydraulic gradient of the rock and soil, piping is easy to occur. When the ratio of the permeability coefficients of the two contacting rock soil layers is greater than 2, piping is likely to occur.

(3) Some engineering cases

1) Ground collapse caused by quicksand

At 4 am on July 1st, 2003, water seepage was found in the Puxi connecting passage of the Shanghai Metro Line 4 section tunnel under construction, and then a large amount of quicksand influxed, causing substantial ground subsidence. At about 9 o'clock in the morning, an eight-story building on the ground surface tilted, and its podium partially collapsed. Due to the timely call to the police, all personnel have been evacuated in advance, so there was no casualty, and all the citizens in the surrounding buildings affected have been evacuated as well (Fig. 6-23).

(a) Podium collapsed

(b) Linjiang garden building (inclination 10°)

Fig. 6-23 Damage caused by quicksand

2) The dam failure of Teton Dam (Fig. 6-24a) in the United States

General situation: earth dam; 90 m heigh; 1000 m long; finished in 1975; destroyed in June 1976.

Reason: seepage failure

Loss: $ 80,000,000; 14 persons died.

Process:

10:30, 5th, June 1976, the seepage with clay and soil occurs on the downstream surface of the dam (Fig. 6-24b).

11:00, the cave enlarges and becomes close to dam top, while the mudflow rate increases (Fig. 6-24c).

Fig. 6-24 The dam failure process of Teton Dam in the United States

11:30, the cave enlarges continuously and the dam foundation has been destroyed. Another seepage cave occurs above the original main cave. The mudflow begins damaging facilities at the dam toe (Fig. 6-24d).

11:50, the cave enlargement accelerates and the damaging erosion on facilities at the dam toe becomes more intensively. Two caves connect (Fig. 6-24e)

11:57, the dam top collapses and the mudflow slumps down quickly (Fig. 6-24f). The collapsed entrance becomes wider. Almost all of the facilities are destroyed absolutely. (Fig. 6-24g).

6.5.3 Floating Effect

Groundwater not only produces buoyancy for the basement and foundation of the building, but also for the rock and soil below the water table. Therefore, the *Code for Design of Building Foundation* (GB 50007—2011) stipulates that when determining the characteristic value of the foundation bearing capacity, both the natural weight of the soil below the bottom of the foundation and the weighted average weight of the soil above the bottom of the foundation and the groundwater level and below shall be taken buoyant unit weight.

(a) Column cracks in an underground garage   (b) In 2020, multiple load-bearing columns cracked at Vanke Sky City in Nanchang, China

Fig. 6-25 Column cracks caused by groundwater floating effect

The heavy rainfall in the rainy season has caused the groundwater level to rise, and the floating effect of groundwater has caused the frequent bursting of underground garage pillars, which should be paid attention to (Fig. 6-25).

6.5.4 Inrushing of Foundation Pit

When there is a confined aquifer at the bottom of a deep foundation pit, excavating the foundation pit will reduce the thickness of the overlying impermeable layer. If the thickness of the impermeable layer is reduced to a certain extent, the head pressure of the confined water can crack or wash away the bottom of the foundation pit, causing inrushing. The sudden surge of the foundation pit will destroy the strength of the foundation and bring great difficulties to the construction. Therefore, during the construction of foundation pits, the possibility of foundation pit surge must be analyzed. In engineering practice, it is usually required that the self-weight stress of the retaining soil layer over the confined water is greater than the head pressure of the confined water. If not, artificial measures must be taken to extract groundwater from the confined aquifer to lower the confined water head to the head pressure that the foundation pit can withstand.

Inrush of foundation pit refers to the engineering geological phenomenon that when the thickness of confined water and water-proof roof at the bottom of foundation pit becomes thinner due to excavation of foundation pit, and is not strong enough to resist the pressure of confined water head, the pressure of confined water head will break through the foundation pit floor.

In order to avoid foundation pit surge, check the safety thickness M of foundation pit bottom, then balance the weight of rock and soil in the range of foundation pit bottom thickness and the pressure of confined head.

$$\gamma M = \gamma_w H \quad (6\text{-}14)$$

γ, γ_w —They are respectively the unit weight of rock and soil at the bottom of foundation pit and the unit weight of groundwater.

H —the confined water head relative to aquifer roof.

M —the thickness of water-proof roof rock and soil after the excavation of the foundation pit. The thickness of the rock and soil layer of the water resisting roof of the confined water should be consistent with the formula (Fig. 6-26):

$$M > \frac{\gamma_w}{\gamma} H \quad (6\text{-}15)$$

In order to prevent the foundation pit

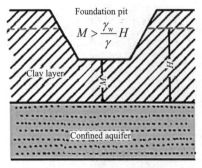

Fig. 6-26 Minimum thickness of water-proof roof at the bottom of foundation pit

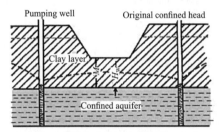

Fig. 6-27 Pumping and depressurization to prevent foundation pit surget

Fig. 6-28 Water-inrush of foundation pit

from bursting, it is necessary to pre-drain the confined aquifer so that its confined water head drops to that the foundation pit can withstand (Fig. 6-27, Fig. 6-28). The pressure head H_w on the roof of the aquifer must be consistent with the following formula:

$$H_w < \frac{\gamma}{\gamma_w} H \qquad (6\text{-}16)$$

6.5.5 Water Gushing and Mud Bursting in Underground Caverns

During the construction of underground caverns, under the action of certain water pressure, along the permeable rock mass (belt) and the cave filled with no (little) sediment, the phenomenon that a large amount of water suddenly occurs is called water gushing. There are three situations that may have serious water gushing problems: ① Underground caverns pass through water-rich layers or other water catchment structures; ② Underground caverns pass through water-rich fault zones, dense joints or other structural fracture zones; ③ Underground caverns pass through karst channels such as water-filled karst caves and underground rivers.

The characteristics of the water gushing from underground caverns in karst areas include: ① The water gushing volume is large and the water pressure is high. When tunnels (especially deep-buried tunnels) penetrate karst pipelines and dissolution cracks connected downward by underground rivers suspended high above the roof of the caves, large flow and high-pressure water gushing often occurs. ② Seasonal. ③ Inhomogeneity. ④ Suddenness.

Fig. 6-29 shows water gushing in the tunnel.

During the construction of underground caverns, under the action of certain water pressure, along the loose (soft) rock belt or the filled karst cave, the sudden gushing of a large amount of water, mud, sand and other mixtures is called mud bursting (Fig. 6-30). The possible mud outburst problems are as follows: ① The underground cavern is located in the loose aquifer. ② Underground caverns pass through the water-saturated fault fractures

Fig. 6-29 Water gushing in the tunnel

Fig. 6-30 Mud bursting

zone or other structural fracture zones. ③ Underground caverns pass through filling-type karst caves, underground rivers and other sections.

6.5.6 Erosion of Groundwater

(1) Corrosion to steel components

The acid corrosion of groundwater on various steel components (such as steel bars, steel pipes and various steel) is mainly related to the acidity of the water. The lower the pH value, the stronger the acidity, and the stronger the corrosive effect of water on metals.

$$Fe+2H^+ \to Fe^{2+}+H_2$$

(2) Corrosion hazards

When the content of some components of groundwater is too much, it will have corrosion hazards to concrete, soluble stones, pipelines, steel and so on. According to the evaluation criteria for the corrosion of building structure materials by groundwater, the types of corrosion can be divided into decomposition corrosion, crystallization corrosion and composite corrosion by crystallization and decomposition.

1) Decomposition corrosion

This mainly refers to the carbonic acid erosion of concrete by groundwater, which is caused by corrosive CO_2 in the water. As we all know, groundwater contains CO_2 and HCO_3^-. CO_2 reacts with $Ca(OH)_2$ in concrete to form calcium carbonate precipitation.

$$Ca(OH)_2+CO_2 \to CaCO_3\downarrow+H_2O$$

Because $CaCO_3$ is insoluble in water, it can fill the voids of concrete and form a protective film around it to prevent the decomposition of $Ca(OH)_2$. However, if the content of CO_2 in the local sewage exceeds a certain value, and the content of HCO_3^- is too low, the excess CO_2 will react with $CaCO_3$ to form $Ca(HCO_3)_2$ and dissolve in water. The reaction formula is as follows.

$$CaCO_3+CO_2+H_2O \leftrightarrow 2HCO_3^-+Ca^{2+}$$

2) Crystalline corrosion

① When excessive SO_4^{2-} is contained in groundwater, it will interact with $Ca(OH)_2$ in concrete to form gypsum dihydrate crystalline $CaSO_4 \cdot 2H_2O$.

$$CaSO_4+2H_2O \leftrightarrow CaSO_4 \cdot 2H_2O$$

When the anhydrite becomes $CaSO_4 \cdot 2H_2O$, its volume will increase by 31%, resulting in an expansion pressure of 0.15 MPa, which will destroy the concrete.

② Calcium sulfate dihydrate reacts with calcium aluminate hydrate to form calcium sulfoaluminate hydrate (also known as cement bacillus). Calcium sulfoaluminate hydrate is combined with a lot of water of hydration, and its volume can expand nearly 2.5 times, which is very destructive.

$$3CaO \cdot Al_2O_3 \cdot 6H_2O+3CaSO_4+25H_2O= 3CaO \cdot Al_2O_3 \cdot 3CaSO_4 \cdot 31H_2O$$

3) Composite corrosion by crystallization and decomposition

When the content of NH_4^+, NO_3^-, Cl^- and Mg^{2+} ions in groundwater exceeds a certain value, they will react with $Ca(OH)_2$ in

Fig. 6-31 Groundwater corrosion hazards to concrete

concrete, for example:

$MgSO_4+Ca(OH)_2 \rightarrow Mg(OH)_2+CaSO_4$

$MgCl_2+Ca(OH)_2 \rightarrow Mg(OH)_2+CaCl_2$

Fig. 6-31 shows groundwater corrosion hazards to concrete.

(3) Corrosiveness evaluation criteria

1) Evaluation of the corrosivity of water and soil to concrete structures by environment type (Table 6-7, Table 6-8)

Affected by the permeability of strata, the evaluation of the corrosion of water and soil on concrete structures should meet the requirements of Table 6-9.

When the corrosion grades evaluated according to Table 6-8 and Table 6-9 are different, comprehensive evaluation should be made according to the following regulations: ① In determining the corrosion grade, it should be evaluated as weak corrosion when only weak corrosion occurs with no medium or strong corrosion; ② In determining the corrosion grade, when there is no strong corrosion, and the highest corrosion level is medium, it should be evaluated as medium corrosion; ③ In determining the corrosion grade, when there are over one strong corrosion, it should be evaluated as strong corrosion.

Environmental Type Classification   Table 6-7

Types	Environmental geological conditions of the site
I	Alpine area, the arid area adjacent water directly; Groundwater in strong permeable layers in alpine and arid areas
II	Groundwater in weak permeable layers in alpine and arid areas; The wet areas of the wet and very wet impermeable aquifers in all climatic regions are directly adjacent water; Groundwater in the strong permeable layer in the humid zone
III	A slightly wet impermeable water layer in various climate zones; Strong permeable layer above groundwater level in various climate zones

Note: 1. Alpine zone refers to an area with an altitude equal to or greater than 3000 m; arid zone refers to an area with an altitude less than 3000 m and a dryness index K value equal to or greater than 1.5; a humid zone refers to an area with a dryness index K value less than 1.5.

2. Strongly permeable layer refers to gravel soil and sandy soil; weakly permeable layer refers to silt and cohesive soil.

3. The soil layer with water content $w<3\%$ can be regarded as dry soil layer without corrosive environmental conditions.

4. When the concrete structure is exposed to the atmosphere on one side while being in contact with surface water or groundwater on the other side, and the water can evaporate on the side exposed to the atmosphere through penetration or capillary action, it should be classified as Class.

5. When there is regional experience, the type of environment can be divided according to regional experience. When there are two types of environments in the same area, they should be selected according to the specific situation.

Evaluation of the corrosiveness of water and soil to concrete structures according to environmental types   Table 6-8

Corrosion grades	Corrosive medium	Environmental Types		
		I	II	III
Micro corrosion Weak corrosion Moderate corrosion Strong corrosion	Sulfate content SO_4^{2-} (mg/L)	<200 200~500 500~1500 >1500	<200 300~1500 1500~3000 >3000	<200 500~3000 3000~6000 >6000
Micro corrosion Weak corrosion Moderate corrosion Strong corrosion	Magnesium salt content Mg^{2+} (mg/L)	<1000 1000~2000 2000~3000 >3000	<2000 2000~3000 3000~4000 >4000	<3000 3000~4000 4000~5000 >5000
Micro corrosion Weak corrosion Moderate corrosion Strong corrosion	Ammonium content NH_4^+ (mg/L)	<100 100~500 500~800 >800	<500 500~800 800~1000 >1000	<800 800~1000 1000~1500 >1500
Micro corrosion Weak corrosion Moderate corrosion Strong corrosion	Caustic alkali content OH^- (mg/L)	<35000 35000~43000 43000~57000 >57000	<43000 43000~57000 57000~70000 >70000	<57000 57000~70000 70000~100000 >100000
Micro corrosion Weak corrosion Moderate corrosion Strong corrosion	Total salinity (mg/L)	<10000 10000~20000 20000~50000 >50000	<20000 20000~50000 50000~60000 >60000	<50000 50000~60000 60000~70000 >70000

Note: 1. The values in the table are applicable to the conditions with alternating wet and dry effects. When there is no alternate dry and wet effects in Category I and II corrosive environments, the sulfate content in the table should be multiplied by a factor of 1.3.

2. The values in the table are applicable to the evaluation of the corrosivity of water. The evaluation of the corrosiveness of soil should be multiplied by a factor of 1.5; the unit is expressed in mg/kg;

3. The caustic alkali (OH^-) content (mg/L) in the table should be the OH^- content (mg/L) in NaOH and KOH.

Evaluation of the corrosivity of concrete structures by ground permeability water and soil   Table 6-9

Corrosion grade	pH		Corrosive CO_2 (mg/L)		HCO_3^- (mmol/L)	
	A	B	A	B	A	B
Micro corrosion	>6.5	>5.0	<15	<30	>1.0	—
Weak corrosion	5.0~6.5	4.0~5.0	15~30	30~60	1.0~0.5	—
Moderate corrosion	4.0~5.0	3.5~4.0	30~60	60~100	<0.5	—
Strong corrosion	<4.0	<3.5	>60	—	—	—

Note:1. In the table, A refers to groundwater directly adjacent to water or in a strongly permeable layer; B refers to groundwater in a weakly permeable layer. Strongly permeable layer refers to gravel soil and sand, and weakly permeable layer refers to silt and cohesive soil.

2. HCO_3^- content refers to the corrosiveness of HCO_3^- of this type of water is soft water who's the salinity of is lower than 0.1 g/L.

3. Take only pH value index into consideration in the evaluation of soil corrosivity. When evaluating its corrosivity, A refers to the strongly permeable soil layer; B refers to the weakly permeable soil layer.

2) Evaluation of corrosion of steel bars in reinforced concrete structures by water and soil (Table 6-10)

3) Corrosion evaluation of soil on steel structure (Table 6-11)

Evaluation of the corrosion of steel bars in reinforced concrete structures　　Table 6-10

Corrosion grades	Cl^- content in water (mg/L)		Cl^- content in soil (mg/L)	
	Long term immersion	alternation of wetting and drying	A	B
Micro corrosion	<10000	<100	—	—
Weak corrosion	10000~20000	100~500	400~750	250~500
Moderate corrosion	—	500~5000	750~7500	500~5000
Strong corrosion	—	>5000	>7500	>5000

Note: A refers to gravel soil, sandy soil, slightly wet silt, hard and hard plastic clay above the groundwater table; B is wet, very wet silt, plastic, soft plastic, and flow plastic clay.

Corrosion evaluation of soil on steel structure　　Table 6-11

Corrosion grades	pH	Oxidation-reduction potential (mV)	Apparent resistivity ($\Omega \cdot m$)	Polarization current density (mA/cm^2)	Mass loss (g)
Micro-corrosion	>5.5	>400	>100	<0.02	<1
Weak corrosion	5.5~4.5	400~200	100~50	0.02~0.05	1~2
Moderate corrosion	4.5~3.5	200~100	50~20	0.05~0.20	2~3
Strong corrosion	<3.5	<100	<20	>0.20	>3

Note: For the evaluation of soil corrosion to steel structures, take the highest corrosion level among the various indicators.

Quiz

1. What are the hydraulic properties of water in rock and soil?

2. How to understand the relativity of aquifer and aquiclude?

3. What are the three types of groundwater according to burial conditions? What are their characteristics?

4. According to the pore properties of aquifers, which three types of groundwater can be divided into? What are their characteristics?

5. What are the physical properties of groundwater? What are the main chemical components in groundwater? What are the chemical properties of groundwater?

6. What is seepage flow? What is the content, principle and application scope of Darcy's law? How to determine the permeability coefficient?

7. What are the types of groundwater corrosion on concrete structures? What are the reasons for the severe corrosion of steel bars by groundwater? What are the evaluation criteria for the corrosion of groundwater on concrete and steel bars? What protective measures can we take to protect reinforced concrete from corrosion?

8. What are the common types of damage in seepage? What is their damage mechanisms?

9. What is inrushing of foundation pit?

Evaluation of the corrosiveness of water and soil to concrete structures according to environmental types Table 6-8

Corrosion grades	Corrosive medium	Environmental Types		
		I	II	III
Micro corrosion	Sulfate content SO_4^{2-} (mg/L)	<200	<200	<200
Weak corrosion		200~500	300~1500	500~3000
Moderate corrosion		500~1500	1500~3000	3000~6000
Strong corrosion		>1500	>3000	>6000
Micro corrosion	Magnesium salt content Mg^{2+} (mg/L)	<1000	<2000	<3000
Weak corrosion		1000~2000	2000~3000	3000~4000
Moderate corrosion		2000~3000	3000~4000	4000~5000
Strong corrosion		>3000	>4000	>5000
Micro corrosion	Ammonium content NH_4^+ (mg/L)	<100	<500	<800
Weak corrosion		100~500	500~800	800~1000
Moderate corrosion		500~800	800~1000	1000~1500
Strong corrosion		>800	>1000	>1500
Micro corrosion	Caustic alkali content OH^- (mg/L)	<35000	<43000	<57000
Weak corrosion		35000~43000	43000~57000	57000~70000
Moderate corrosion		43000~57000	57000~70000	70000~100000
Strong corrosion		>57000	>70000	>100000
Micro corrosion	Total salinity (mg/L)	<10000	<20000	<50000
Weak corrosion		10000~20000	20000~50000	50000~60000
Moderate corrosion		20000~50000	50000~60000	60000~70000
Strong corrosion		>50000	>60000	>70000

Note: 1. The values in the table are applicable to the conditions with alternating wet and dry effects. When there is no alternate dry and wet effects in Category I and II corrosive environments, the sulfate content in the table should be multiplied by a factor of 1.3.

2. The values in the table are applicable to the evaluation of the corrosivity of water. The evaluation of the corrosiveness of soil should be multiplied by a factor of 1.5; the unit is expressed in mg/kg;

3. The caustic alkali (OH⁻) content (mg/L) in the table should be the OH⁻ content (mg/L) in NaOH and KOH.

Evaluation of the corrosivity of concrete structures by ground permeability water and soil Table 6-9

Corrosion grade	pH		Corrosive CO_2 (mg/L)		HCO_3^- (mmol/L)	
	A	B	A	B	A	B
Micro corrosion	>6.5	>5.0	<15	<30	>1.0	—
Weak corrosion	5.0~6.5	4.0~5.0	15~30	30~60	1.0~0.5	—
Moderate corrosion	4.0~5.0	3.5~4.0	30~60	60~100	<0.5	—
Strong corrosion	<4.0	<3.5	>60	—	—	—

Note:1. In the table, A refers to groundwater directly adjacent to water or in a strongly permeable layer; B refers to groundwater in a weakly permeable layer. Strongly permeable layer refers to gravel soil and sand, and weakly permeable layer refers to silt and cohesive soil.

2. HCO_3^- content refers to the corrosiveness of HCO_3^- of this type of water is soft water who's the salinity of is lower than 0.1 g/L.

3. Take only pH value index into consideration in the evaluation of soil corrosivity. When evaluating its corrosivity, A refers to the strongly permeable soil layer; B refers to the weakly permeable soil layer.

2) Evaluation of corrosion of steel bars in reinforced concrete structures by water and soil (Table 6-10)

3) Corrosion evaluation of soil on steel structure (Table 6-11)

Evaluation of the corrosion of steel bars in reinforced concrete structures   Table 6-10

Corrosion grades	Cl^- content in water (mg/L)		Cl^- content in soil (mg/L)	
	Long term immersion	alternation of wetting and drying	A	B
Micro corrosion	<10000	<100	—	—
Weak corrosion	10000~20000	100~500	400~750	250~500
Moderate corrosion	—	500~5000	750~7500	500~5000
Strong corrosion	—	>5000	>7500	>5000

Note: A refers to gravel soil, sandy soil, slightly wet silt, hard and hard plastic clay above the groundwater table; B is wet, very wet silt, plastic, soft plastic, and flow plastic clay.

Corrosion evaluation of soil on steel structure   Table 6-11

Corrosion grades	pH	Oxidation-reduction potential (mV)	Apparent resistivity ($\Omega \cdot m$)	Polarization current density (mA/cm^2)	Mass loss (g)
Micro-corrosion	>5.5	>400	>100	<0.02	<1
Weak corrosion	5.5~4.5	400~200	100~50	0.02~0.05	1~2
Moderate corrosion	4.5~3.5	200~100	50~20	0.05~0.20	2~3
Strong corrosion	<3.5	<100	<20	>0.20	>3

Note: For the evaluation of soil corrosion to steel structures, take the highest corrosion level among the various indicators.

Quiz

1. What are the hydraulic properties of water in rock and soil?

2. How to understand the relativity of aquifer and aquiclude?

3. What are the three types of groundwater according to burial conditions? What are their characteristics?

4. According to the pore properties of aquifers, which three types of groundwater can be divided into? What are their characteristics?

5. What are the physical properties of groundwater? What are the main chemical components in groundwater? What are the chemical properties of groundwater?

6. What is seepage flow? What is the content, principle and application scope of Darcy's law? How to determine the permeability coefficient?

7. What are the types of groundwater corrosion on concrete structures? What are the reasons for the severe corrosion of steel bars by groundwater? What are the evaluation criteria for the corrosion of groundwater on concrete and steel bars? What protective measures can we take to protect reinforced concrete from corrosion?

8. What are the common types of damage in seepage? What is their damage mechanisms?

9. What is inrushing of foundation pit?

What are the conditions for inrushing of foundation pit? How to prevent inrushing of foundation pit?

10. What is quicksand? What are the destructive effects of quicksand? What are the conditions for the formation of quicksand? What measures can we take to prevent quicksand?

11. What is piping? What are the destructive effects of piping? What are the conditions for piping? What measures can we take to prevent piping?

12. What are the methods for artificially lowering the groundwater level?

Chapter 7
Adverse Geological Phenomena and Geological Problems

7.1 Earthquake

7.1.1 Basic Conceptions

An earthquake is the tectonic movement of the earth's crust caused by the internal forces of the earth. The rock formation in deep underground is vibrated due to the sudden movement of active faults or other reasons, and are transmitted to the surface in the form of elastic waves. In describing the earthquake, the following basic concepts need to be clarified.

The place where an earthquake occurs in the crust or mantle is called earthquake source. The vertical projection of the seismic source on the ground is called the epicenter, which is also called the source area. It is usually a region, but it is often taken as a point in earthquake study. The epicenter can be regarded as the center of vibration on the ground, which can be divided into micro-epicenter and macro-epicenter. The ground vibration near the epicenter is the greatest, and weaken away from the epicenter. Certain range around the epicenter is the epicentral area, the most struck areas during earthquakes.

Under the influence of the same earthquake, the line connecting the points with the same degree of damage on the ground is called isoseismal. According to isoseismal diagrams, the location of the macro-epicenter can be determined. According to the shape of the isoseismal lines in the epicenter area, the trend of the earthquake-producing fault (seismic fault) can be inferred.

The vibration caused by an earthquake propagates from the source to all directions in the form of waves, and those waves are called seismic waves. Seismic waves can be divided into body waves and surface waves. Body waves are divided into longitudinal waves (P-wave) and transverse waves (S-wave), as shown in Fig. 7-1.

Longitudinal waves are also known as P-waves. The particle vibration direction is consistent with the wave's forward direction, and it propagates forward in a dense and sparse manner. Longitudinal waves can propagate in solid, liquid and gaseous states. The propagation speed of longitudinal waves is fast, which is the first wave to reach the surface. It has a short period and small amplitude. The energy of longitudinal waves accounts for about 7% of the energy of seismic waves.

Transverse waves are also known as S-wave. Its particle vibration direction is

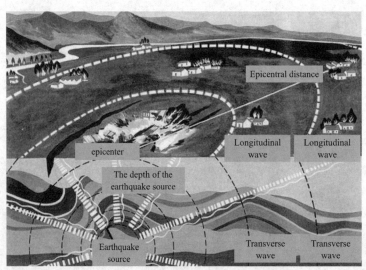

Fig. 7-1 Schematic diagram of earthquake nouns

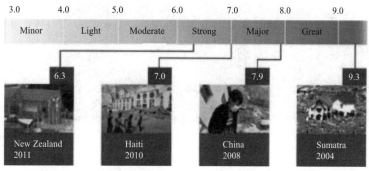

Fig. 7-2 Schematic diagram of Richter scale classification

perpendicular to the wave's forward direction. The medium does not change in volume during propagation, but the change in shape with a longer period, a large amplitude. Transverse waves are the second wave to the surface, and its energy accounts for about 26% of the total seismic wave energy.

Surface wave (also divided into Rayleigh wave, R-wave and Luff wave, L-wave) is a secondary wave motivated after the body wave reaches the ground. It only propagates on the surface and disappears quickly below the ground. The surface wave has a large wavelength, large amplitude, and large energy, accounting for about 67% of the total energy of the seismic wave.

When an earthquake occurs, the first thing to reach the ground buildings is always the longitudinal wave, and people first feel the up and down vibration; secondly, the transverse wave by which people feel the shaking from side to side; the surface wave arrives last. When transverse waves and surface waves arrive, the ground shakes the strongest and buildings are most severely destroyed.

The magnitude of an earthquake is an index to measure the amount of energy released by the seismic source. Earthquakes of magnitude 7 or higher are called strong earthquakes. At present, the general international seismic standard is Richter scale (Fig. 7-2), dividing earthquake into 10 grades.

Earthquake intensity is an indicator that reflects the actual impact and damage degree on a specific location during an earthquake. And the seismic intensity is not only related to the magnitude, but also related to various factors such as the depth of the focal point, the distance from the epicenter, and the seismic performance of building structures. In general, the higher the magnitude, the shallower the focal point, and the closer to the epicenter, the higher the seismic intensity and the greater the damage. At present, the commonly used earthquake intensity standard is Intensity scale (Table 7-1), dividing earthquake into 12 grades.

Classification of earthquake intensity scale Table 7-1

Mercalli earthquake intensity (and peak acceleration)			
I	Not felt	VII	Slight damaged or local damaged (~0.1 g)
II	Felt only by few people at rest indoors	VIII	Masonry damaged
III	Felt only by several people at rest indoors	IX	Foundations damaged
IV	Windows rattle (<0.02 g)	X	Buildings destroyed (>0.6 g)
V	Felt by most people outdoors	XI	Most buildings buckled
VI	People are frightened	XII	Almost all buildings destroyed

(a) Tectonic earthquake (b) Volcanic earthquake

Fig. 7-3 Tectonic earthquake and volcano earthquake

According to causes, earthquakes can be divided into artificial induced earthquakes, tectonic earthquakes, volcano earthquakes and collapse earthquakes.

(1) Artificial induced earthquakes

Artificially induced earthquakes are mainly caused by two conditions. First is reservoir water storage and the like, which can increase the load on the underground rock formations. If there are large cracks or structural fault zones in the ground, flooding can occur on the fault surface and cause earthquakes. Second is huge explosions such as underground nuclear explosions or underground blasting, which can have a strong impact on the ground, and promote the release of tectonic stress in the crust and induce earthquakes.

(2) Tectonic earthquakes

Earthquakes caused by tectonic movements of crustal fractures are called tectonic earthquakes (Fig. 7-3a). Tectonic earthquakes result from stress release and fault dislocation due to tilt, folds, fractures, dislocations, and large-scale magmatic activities in the rock formations that make up the crust caused by crustal movement. Tectonic earthquakes account for about 90% of the total number of earthquakes.

(3) Volcano earthquakes

An earthquake by a volcanic eruption is called a volcanic earthquake (Fig. 7-3b). This type of earthquake is relatively strong, but the earthquake area is relatively small. It only accounts for about 7% of the total number of earthquakes.

(4) Collapse earthquakes

Earthquakes caused by strata collapse, landslides, giant landslides, etc. are called subsidence earthquakes. Such type of earthquakes occurs when the strata collapse occurs in the limestone karst area where the karst dissolution causes the cavern to expand continuously, and lead to the collapse of the overlying strata. Generally, a collapse earthquake has a small seismic energy and a small impact area. Such earthquakes account for only about 3% of the total.

7.1.2 Earthquake Destructive Effects and Seismic Fortification Measures

The destructive effects of earthquakes can be divided into two aspects: vibration damage and ground damage. Common ground damages include ground fracture, building destruction, landslides, foundation effects. The main damage modes of the Sumatra earthquake are resonance damage, standing wave damage, phase difference damage, seismic liquefaction.

(1) Resonance destruction

During an earthquake, the seismic waves emitted from the seismic source travel through the soil layer and pass-through multiple reflections of interfaces of different natures, resulting in seismic waves of different periods. If the seismic wave of a certain period is close to the natural period of the foundation soil, the amplitude of this seismic wave will be amplified due to the effect of resonance. This period is called the predominant period T.

The predominant period can be calculated by the following formula:

$$T = \sum_{i=1}^{n} \frac{4h_i}{V_s} \quad (7\text{-}1)$$

Where h_i is the thickness of the i-th layer, which is generally calculated up to the bedrock; V_s is the shear wave velocity. According to earthquake statistics, depending on the degree of softness and hardness of the foundation soil, the predominant period can be divided into four levels:

I——Stable rock formation, the predominant period is 0.1~0.2 s, with an average of 0.15 s;

II——General soil layers, the predominant period is 0.2~0.4 s, with an average of 0.27 s;

III——Soft soil layers, the predominant period is between II and IV;

IV——Abnormally loose soft soil layers, the predominant period is 0.3~0.7 s, with an average of 0.5 s.

(2) Standing wave destruction

During an earthquake, when two seismic waves with the same amplitude and frequency phase but moving directions move toward the same direction and converge, a standing wave is formed with a doubled amplitude. When affected by standing waves, building will be severely damaged, which is standing wave damage. When the seismic wave under the same conditions meets the seismic wave reflected from the valley in a certain place, it will cause standing wave damage to the buildings in that place.

(3) Phase differential destruction

When the length of the building is less than the ground vibration wavelength, there will be harmonic vibration with uniform amplitude in the building and the foundation. When the length of the building is close to or greater than the site vibration wavelength, the vibration phases of the building and foundation are inconsistent to form a very uncoordinated vibration. At this time, no matter how large the ground vibration displacement (amplitude) is, the average amplitude of the building is zero. In this case, the vibration of the foundation violently hits the underground structure of the building and causes damage at its weakest part. Such damage is the phase differential destruction.

(4) Earthquake liquefaction and subsidence

For saturated silt, during the earthquake, the vibration will cause the pore water pressure in the saturated soil layer to rise suddenly. The pore water pressure can't dissipate immediately, and effective pressure between the sand grains will be reduced. If the effective pressure disappeared, the sand layer will completely lose shear strength and load carrying capacity, and turns into liquid, which is the so-called sand liquefaction. The macroscopic manifestations of seismic liquefaction include water spraying and sand liquefaction, and underground sand liquefaction. Earthquake liquefaction will cause surface subsidence and deformation, called earthquake subsidence, causing deformation and damage of ground structures.

(5) Earthquake induced geological hazard effects

The large amount of loose ground matter produced by earthquakes can easily cause mudslide. Strong earthquake can also stimulate the rock and soil mass on the slope loosening and instability, resulting in landslide and collapse and other undesirable geological hazards. It can destroy houses and roads, even bury villages and block rivers. Therefore, construction sites and main lines should avoid areas that may be affected by earthquakes and trigger geological disasters.

The basic seismic conceptual design is as follows:

1) Basic intensity: The intensity of common site met in the future 100 years with the exceedance probability of 10%.

2) Site intensity: the maximum intensity which could be met in the effective using period for construction site. Generally, it should be one degree or half degree higher

 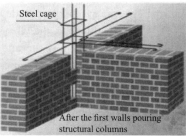 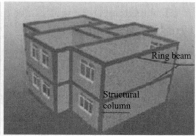

Fig. 7-4 Seismic reinforcement

than basic intensity.

3) Design intensity: adjusted intensity during seismic design according to significance, seismic properties of buildings and the economic condition, and so on.

The standard for seismic design of buildings in China is that no damage happens in small earthquake, damage under moderate earthquake is repairable, no collapsing in the strong earthquake. The conventional shock absorption and isolation design is as follows.

1) Select a favorable site: The selection should meet the requirement of projects based on well knowledge of seismic activities, engineering geological conditions and seismic geological conditions.

2) Anti-liquefaction, anti-subsidence, anti-sliding measures: Including grouting, replacement, piling, drainage consolidation.

3) Foundation design: Increase the foundation buried depth, and select the better foundation type.

4) Multi-story masonry structure buildings: The damage is concentrated at wall under strong earthquake, while the damage of floor is lighter.

Methods of seismic reinforcement (Fig. 7-4) are as follows. ① Set up reinforced concrete columns to reduce the damage of the wall body. ② Add reinforced concrete ring beams connected with structural columns, enhancing the integrity of buildings. ③ Strengthen the connection of the wall, the floor and beam should have enough length and reliable connection. ④ Strengthen the integrity of the staircase.

5) Frame structure buildings

The damage is concentrated at frame beam column joint and filling wall. Methods of seismic strengthening including: ① The framework should be designed as a ductile frame. ② Column, node, anchoring should be strong. ③ Control the minimum reinforcement ratio and the minimum reinforcement diameter. ④ Set up seismic joint.

7.2 Landslides

7.2.1 Basic Conceptions

Landslide refers to the deformation and failure of the slope rock and soil to slid downwards and outwards along the weak geological surface under the action of gravity and other factors. A typical full-grown landslide has the following basic structural features (Fig. 7-5), which is an important sign for identifying and judging a landslide.

The features of landslide are as follows: (1) Sliding body: the whole sliding part of landslide. (2) Sliding cliff: the boundary surface between the back edge of sliding body and unmoved part of the mountain. (3) Sliding surface: the boundary plane between sliding body and unmoved underlying rock and soil. (4) Sliding zone: cracked zones caused

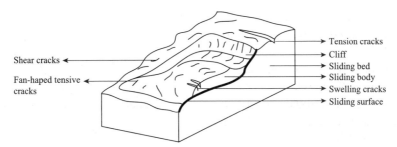

Fig. 7-5 Three-dimensional sketch (draft) of landslide

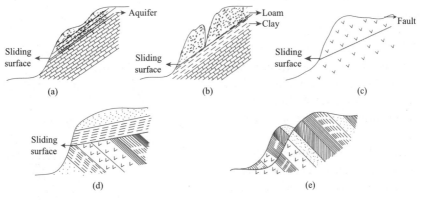

Fig. 7-6 Shapes of sliding surface

by landslide near sliding surface, and the structural disturbance zones which are several centimeters to several meters thick. (5) Sliding bed: the unmoved underlying rock and soil when sliding body moves down. (6) Landslide tongue: the convex part of landslide front like a tongue. (7) Landslide step: A stepped ground formed due to the sliding speed difference of all kinds of rock and soil. (8) Tension cracks: generally, on the back platform, like arc shape. (9) Swelling cracks: the tension cracks when the front part of landslide is blocked then bumping. (10) Fan-shaped cracks: distributing on the landslide tongue like fans. As shown in Fig. 7-6.

7.2.2 Classification of Landslides

In order to understand landslides and treat disastrous landslides, it is necessary to classify the landslides. Landslides in China are classified as follows in *Standard of Classification for Geological Hazards* (T/CAGHP 001—2018).

(1) According to mechanical conditions

Retrogressive landslide (Fig. 7-7a): The lower part of the sliding body loses the balance to slide firstly, then developing upwards, until the upper part of the sliding body is driven by the traction and follows the sliding.

Pushing style landslide (Fig. 7-7b): The local failure firstly happens on the upper part of the sliding body, and then the sliding surface

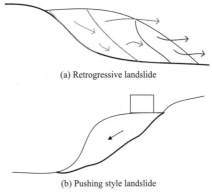

Fig. 7-7 Mechanical classification of landslides

further links up and compresses the lower sliding body slowly until the whole sliding body slides lastly.

(2) According to the relationship between the sliding surface and rock layer

Landslides can be divided into three types: homogeneous landslide, bedding landslide and cut-layer landslide.

1) Homogeneous landslides occur in homogeneous rock and soil without obvious bedding; the landslide surface is generally arc-shaped, as shown in Fig. 7-8(a).

2) When the rock layer inclination is consistent with the slope and the dip angle is smaller than the slope angle, it often slides along the weak structural surface between the layers to form a landslide, as shown in Fig. 7-8(b).

3) Sliding along the base-cover interface, and the upper overburden (slope layer, residual layer and artificial accumulation, etc.) slides along the underlying bedrock surface, as shown in Fig. 7-8(c).

4) Most of cut-layer landslides occur in a group or two groups of joint surfaces outside the inclined slope to form a through sliding surface, as shown in Fig. 7-8(d).

(3) According to the material composition of the landslide body

Landslides can be divided into two types: soil landslides and rock landslides.

Soil landslides: This category mainly refers to landslides in which the landslide material is mainly composed of soil or loose accumulation materials. The landslide has obvious landforms, slow sliding speed. The slope of the sliding bed is relatively slow, the scale is small, and the sliding more appear in groups. It can be divided into two categories: coarse-grained soil landslides and fine-grained soil landslides according to the particle size and material composition of the landslide.

Rock landslide: It mainly composed of various rock mass with joints and fissures in the rock mass.

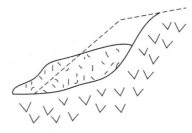
(a) Homogeneous landslide

(b) Bedding landslide

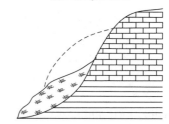
(c) Bedding landslide at the interface between slope-accumulation and bedrock

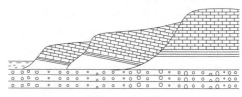
(d) Cut-layer landslide

Fig. 7-8 Relationship between slide surface and geological structure

7.2.3 Formation Conditions of Landslides

The root causes of landslide formation are the changes in the nature of the slope rock and soil, the internal structure of the slope medium, and the spatial form of the slope. Factors such as water, earthquakes and human activities can change the shape of the slope, deteriorate the properties of the rock and soil of slope, and add additional loads, and finally lead to the attitude of landslides.

(1) Strata and lithology

In hard rock strata, landslides are less likely to occur. But when there are weak fractured zones or thin weathered layers in the rock mass, steep dip angles and groundwater activity, the rock strata may slide along the weak surface. Landslides mainly occur in soil layers that are easy to be hydrophilic and softened, and some soft rocks, such as clay soil, loess, weathered rock, and soil layers that are easy to swell and soften in contact with water. Soft rocks include shale, mudstone and marl, phyllite and weathered tuff.

(2) Geological structure

The internal geological structures of the slope (such as rock or soil layers, joints, cracks), are often decisive factors affecting the stability of the slope. Landslides often occur in the following situations:

1) There are thin layers of soft rock, weak fracture zones or thin weathered layers in the hard rock strata. When the inclination of the weak interlayer is steep and there is groundwater activity, the rock strata may slide along the weak interlayer.

2) When the slope has layered rocks such as basalt, landslide is easy to happen along the bedding, the coal-bearing strata are prone to landslide along the coal seams.

3) Landslides are more likely to occur in the structural planes of metamorphic rocks. When atmospheric precipitation seeps down along the voids of the soil, it is easy to form a weak saturated soil layer on the underlying bedrock surface, causing the soil to slide along this weak surface.

4) Landslides tend to occur along slope body with fault fracture zones and dense zones of joint fracture.

(3) Climatic conditions

Hot and dry summers cause the clay layer to crack. When heavy rain occurs, water can seep into the soil along cracks (landslides) causing, landslide will occur against such situation.

(4) Topography and geomorphic conditions

The slope height, inclination angle and surface undulation shape of the slope have a great influence on its stability. The smoother the slope angle and the lower the slope height, the better the stability of the slope. The undulating hilly landforms, the margins of mountain basins and plain landforms are all geomorphic units where landslides are easy to occur.

(5) Water

Water can soften the rock and soil, reduce their strength, and accelerate the weathering of the rock and soil. Relevant data show that more than 90% of landslides is related to the action of water. The role of water is manifested in the following aspects:

1) Atmospheric precipitation seeps down along the surface of the soil slope, increasing the weight of the soil layer, changing the original stress state of the soil slope, and may cause the soil slope to slide.

2) The infiltration of water causes changes in the mechanical properties of the soil slope medium and leads to the landslides. In the upper part of the slope, joint cracks are developed and weathering is strong, forming an aquifer. When the lower part of the slope is complete or relatively water-isolated, landslides are likely to occur along the interface between the aquifer and water-isolated layer in rainy season.

3) The existence of the fault zone leads to the hydraulic among between groundwater, surface water and different aquifers. The change of the stress state of the slope body caused by the seepage hydraulic effect in the slope body will cause the slope body to slip.

4) The dissolution, erosion and erosion of the slope medium caused by seepage of groundwater change the internal structure of the slope body and cause the general movement of the slope.

(6) Earthquake

Earthquake will first destroy the structure of rock and soil mass, which can make the silty sand layer liquefaction, so as to reduce the shear strength of rock and soil mass. At the same time, the seismic wave is transmitted in the rock and soil body, making the rock and soil body bear the seismic inertia force, increasing the sliding force of the landslide body and promoting the occurrence of landslide.

(7) Human Factors

Man-made destruction of the surface covering will increase surface water infiltration. In the process of civil construction, improper slope cutting, failure of slope support, and load increasing will damage the stability of the original slope. Leakage from diversion irrigation will allow water to seep into the slope, adding more possibility of sliding.

7.2.4 Developing Process of Landslides

The development process of landslides is usually divided into three stages: creep deformation stage, sliding failure stage and gradually stabilizing stage. Studying the development process of landslides is of great significance for understanding landslides and choosing proper anti-slip measures.

Step 1: Creep deformation stage (t_0~t_2 period in Fig. 7-9, in which t_0~t_1 period is the initial deformation stage)

The slope is usually stable before slippage occurs. Under the action of natural conditions and human factors, the strength of the rock and soil of the slope can be gradually reduced, then reducing the stability of the slope. In part of the slope, the deformation is first caused by small movement, and then the deformation is further developed until intermittent tensile cracks appear on the slope surface. With the emergence of tensile cracks, the water seepage effect is strengthened, the deformation further develops, the trailing edge is stretched, the cracks widen, and small deviations begin to appear, and shear cracks on both sides also appear one after another. The rock and soil near the foot of the slope will be squeezed (Fig. 7-10), and the water will seep near the exit of the landslide. At this time, most of the sliding surface is formed, but not connected. The slope deformation will further develop, the post-tension cracks widen continuously, and the offset increases. The sliding surface is established, and then the whole begins to slide. The period during the stability of the slope is destroyed, and cracks appear on the slope, and the time before the slope starts to slide as a whole, is called the creep deformation stage of the landslide. Creep deformation period can be long or short, as long as several years, or as short as a few months or a few days.

Step 2: Sliding failure stage (t_2~t_3 period in Fig. 7-9)

When the slope slides down as a whole, the trailing edge quickly sinks, and the landslide wall becomes more exposed. The landslide body forms a stepped terrain on the ground, the trees on the landslide are leaning here and there, looks like a "drunken forest" (Fig. 7-11). During the sliding process of the

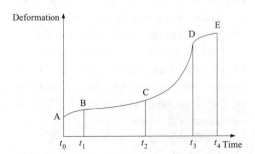

Fig. 7-9 Schematic diagram of typical deformation curve of landslide

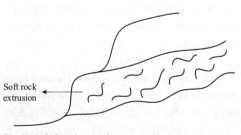

Fig. 7-10 Soft rock extrusion

Fig. 7-11 Drunken-like trees

Fig. 7-12 Sabre-like trees

landslide, the humidity near the sliding surface increases, and the structure of the rock and soil is further damaged due to repeated shearing, which causes the shear strength of the rock and soil to be further reduced, and promotes the acceleration of the sliding of the landslide. Sliding is often accompanied by loud noise and huge waves, which can be a huge disaster for the environment.

Step 3: Gradual stabilization stage ($t_3 \sim t_4$ period in Fig. 7-9)

Because the landslide body has kinetic energy in the process of sliding, the landslide can cross the equilibrium position and slide to a further place. After sliding, the dip angle of the rocks becomes slow or steep, and the azimuth of faults and joints also changes regularly. Under the action of gravity, the loose rock and soil on the landslide body is gradually compacted, and various cracks on the ground surface are gradually filled. The strength of rock and soil near the slide zone increases again due to the compaction and consolidation, which greatly improves the stability of the whole landslide. After some time, the "drunken forest" on the landslide was growing up vertically again, but its lower part could not be straightened. So, it is called "saber trees" (Fig. 7-12), which is a phenomenon of the landslide stabilization.

The rock and soil in the landslide are compacted and there are no obvious cracks on the surface. When there is no water seeping or cool spring water flowing out from the front of the landslide, the landslide has basically stabilized. After the landslide tends to stabilize, if the main factors that caused the landslide have been eliminated, the landslide will no longer slide and turn into long-term stability. On the contrary, if the main factors causing landslides are not completely eliminated, stabilized landslides will eventually slide again.

7.2.5 Field Identification of Landslides

A series of changes in landslides provide us with markers to identify landslides in the field. Its signs mainly include stratigraphic structures, topographic features, hydrogeological signs and landslide precursor phenomena.

(1) Stratigraphic tectonic markers

The integrity of strata in the landslide area is often damaged by sliding, and there is a phenomenon of disturbing loosening. The attitude of rock strata has changed obviously (such as incoherent cracks, and displacement); structural discontinuity is the sign of the existence of landslides.

(2) Topographic features

The presence of landslides often causes slopes to form chair-shaped terrain and trough-shaped terrain. The upper part has steep walls and arc-shaped tensile cracks; the central potholes are undulating. There are bulging mounds in the lower part, protruding outward in a tongue-like shape. Phenomena like "drunk forests", "saber trees", cracked buildings, leanings tend to appear on both sides.

(3) Hydrogeological markers

The original condition of the aquifer in the landslide area is often destroyed without certain rules to follow, and make the landslide a separate body of water. Its hydrogeology has become particularly complex, such as irregular water table, uncertain flow direction, rows of spring water overflow at the bottom of the slope. All of these phenomena can be used as markers to identify landslides.

(4) Identification of landslide precursor phenomena

Different types of landslides with various characteristics will show a variety of abnormal phenomena before sliding, which are as follows:

Before sliding, at the foot of the front slope of the landslide, spring water that had been blocked for many years revived, and the water table of the well suddenly changed, or other similar abnormal phenomena can occur.

At the leading edge of the landslide, the soil and rock fell sporadically. A large number of expanding cracks appeared near the foot of the slope. This is a clear sign that the landslide is pushing forward.

7.2.6 Prevention Measures and Treatments of Landslides

(1) General principles

Prevention and treatment of landslide should be mainly based on prevention and supplemented by treatment. It is necessary to control the developing landslide, prevent the occurrence of possible landslide. The specific prevention and treatment principles can be summarized as follows.

1) On the basis of finding out the engineering geological conditions of the slope deformation and damage area, analyze the main and secondary factors affecting the slope stability, and select corresponding preventive measures.

2) The scale and boundary conditions of slope deformation and failure must be clarified before remediation. Corresponding measures should be taken according to the scale of slope deformation.

3) Take different prevention measures according to the importance of the project. It is necessary to improve the safety and stability of the major projects which can be severely damaged after slope becomes instable. Simpler prevention and control measures can be taken for non-major works and temporary works.

(2) Prevention measures and treatments

Engineering measures for landslide prevention and control can be roughly divided into three categories: drainage, mechanical balance and improving the properties of rock and soil on the sliding surface.

1) Discharging water

The purpose of drainage measures is to reduce the amount of water entering the slide body and reduce the sliding force of the landslide.

Three-dimensional drainage system (Fig. 7-13): including surface water and underground water.

① Surface water: set up drainage trench and catch drain.

② Underground water: Horizontal

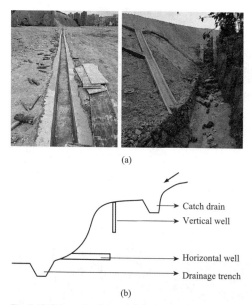

(a)

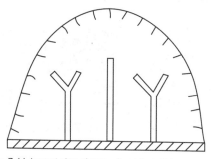

(b)

Fig. 7-13 Schematic diagram of drainage system

(a) Drainage corridor

(b) Corridor entrance

(c) Drainage trench

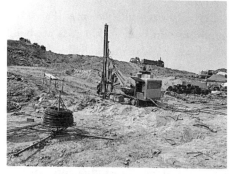

(d) Drainage holes construction

Fig. 7-15 Some pictures of drainage system at construction sites

Fig. 7-14 Layout plan of supporting blind ditch

drainage: set up blind ditch, blind hole; vertical drainage: set up well, drilling.

For the shallow groundwater in the sliding body, the supporting blind ditch, which has the dual functions of drainage and support, is often used to intercept and drain the groundwater. The position of the supporting blind ditch is mostly parallel to the sliding direction, and it is generally set at the place where the groundwater is exposed, and the plane is Y-shaped or I-shaped (Fig. 7-14) The front water surface of the blind ditch is made of a permeable layer, and the back water surface is a water blocking layer to prevent the water collected in the blind ditch from penetrating into the sliding body; the top of the ditch is laid with an impermeable layer. Fig. 7-15

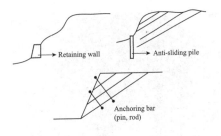

Fig. 7-16 Common retaining structures

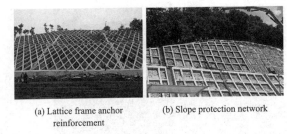

Fig. 7-17 Common support measures

Fig. 7-18 Unloading and cutting diagram

Fig. 7-19 Some pictures of slope cutting

shows some pictures of the construction of drainage system at the project sites.

2) The supporting structure (mechanical balance)

Retaining walls, anti-slide piles (Fig. 7-16) or reinforcement with anchor rods are built in the lower part of the landslide body to increase the anti-sliding force of the lower part of the landslide. Fig. 7-17 shows the common support measures in engineering.

3) Unloading and cutting (Fig. 7-18)

By cutting the slope angle or lowering the slope height to reduce the weight of unstable parts of the slope, the sliding force of the upper part of the landslide will be reduced. Fig. 7-19 shows some pictures of common cutting construction in engineering.

4) Improving the properties of rock or soil at sliding surface or zones

The main purpose of improving the properties and structures of rock and soil mass is to increase their strength of the slope. Such measures include: consolidation grouting for rock landslides; electrochemical reinforcement, freezing, and roasting for soil landslides. Generally speaking, multiple methods are often used based on the formation of the landslide for comprehensive treatment.

7.3 Dangerous Rock and Collapse

7.3.1 Basic Conceptions

Collapse refers to the dynamic geological phenomenon that under the action of gravity, the vertical movement, rolling and jumping along the hillside of the rock and soil body, and finally accumulates at the foot of the slope.

Dangerous rock mass refers to the rock mass that is cut by multiple sets of geological tectonic fractures on a steep slope, has poor stability, and collapse or rock fall may occur in the rock mass (Fig. 7-20). The damage brought by collapse is as follows (Fig. 7-21).

Several cubic meters to hundreds of cubic meters of rock blocks can collapse in small-scale collapses; tens of thousands of cubic meters to tens of millions of cubic meters of rock blocks can collapse in large-scale collapses. The collapse can develop gradually, but happen suddenly, which directly threatens the safety and normal operation of residents, buildings, highways and their corresponding buildings in front of the dangerous rock mass, causing casualties and huge economic losses every year.

7.3.2 Classification of Collapses

The classification of collapses of dangerous rock is based on a systematic study of dangerous rock mass. According to its size, mode of movement, block orientation and instability mode, the common classification is as follows.

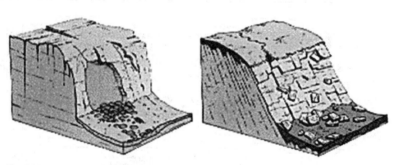

Fig. 7-20 Typical form of collapse: rockfall

Fig. 7-21 Collapse damages

(1) Classification based on substance composition and inducing factors

Standard of Classification for Geological Hazards (T/CAGHP 001—2018) has made the following classification (Table 7-2).

(2) According to movement mode

According to the movement mode after instability, there are five categories: toppling collapse; slip collapse; boom collapse; pull crack collapse and staggered collapse (Table 7-3).

(3) Classification by scale

According to the scale (volume), the collapse of dangerous rock can be divided into: super large collapse (≥ 1 million m^3), large collapse (10^5 to 1 million m^3), medium collapse (10^4 to 10^5 m^3), and small collapse ($<10^4$ m^3).

(4) Classification based on the height difference between the top of the dangerous rock and the foot of the steep cliff

According to this classification, it can

Classification based on substance composition and inducing factors Table 7-2

Classification factors	The types of collapse	Description of its characteristics
Substance composition	Soil collapse	A collapse occurring in soil mass.
	Rock collapse	A collapse occurring in rock mass.
Inducing factors	Natural dynamic collapse	A collapse formed by natural processes such as precipitation, erosion, weathering and denudation, and earthquake.
	Man-made dynamic collapse	A collapse caused by engineering disturbance, blasting and artificial loading.

Classification of collapse based on formation mechanism Table 7-3

Type	Toppling collapse	Slip collapse	Boom collapse	Pull crack collapse	Staggered collapse
Lithology	Strata of loess, free-standing or steep slopes	It's mostly hard and soft rock	Loess, clay, soft rock beneath hard rock	It's more common in hard and soft rock	Hard rock strata, loess
Structural plane	Mostly vertical joints, within a steep slope-self-supporting layer	Structural surface that tends to face the open surface	The upper part is a vertical joint, and the lower part is a nearly horizontal structural plane	Mostly weathered cracks and vertical tension cracks	Vertical fissures are developed, and the structural plane usually with no inclination to air
Landscape	Canyon, independent bank slope, cliff	Steep slopes are usually greater than 55°	Steep slope	The cliff jutting out from above	Steep slopes greater than 45°
Stress state	It is mainly affected by overturning moment	The slip plane is mainly subjected to shear forces	The lower soft rock is pressed vertically	tension	Shear force due to dead weight
Initial mode of movement	Dumping	Slip and fall	Swelling accompanied by sinking slip, tipping	Tearing, falling	Fall

be divided into: ultra-high dangerous rock ($H \geqslant 100$ m); high dangerous rock ($50 \leqslant H < 100$ m); intermediate dangerous rock ($15 \leqslant H < 50$ m); low dangerous rock ($H < 15$ m).

7.3.3 Formation Conditions of Collapse

The destruction of the balance and stability of the slope rock mass is the basic cause of collapse. The main occurrence conditions and development factors of collapse can be divided into the following aspects:

(1) Terrain conditions

The slope and the structural characteristics of hills are necessary conditions for the formation of the collapse. Large-scale collapses generally occur on steep slopes with a height greater than 30 m and a slope angle greater than 55° (mostly between 55° and 75°). Due to stress redistribution and unloading, the front edge of the slope produces long and deep tensile cracks to form a continuous separation plane through combination with other structural planes, and cause collapses under the action of inducing factors. The external shape of the slope also has a certain influence on the formation of collapse. However, the structure of the hillside surface should be in combination with the cracks and weathering of the rock strata as evaluating factors.

(2) Rock properties and structures of rock mass

Hard rock has greater shear strength and weathering resistance to form high and steep slopes. Therefore, collapse often occurs on slopes composed of hard and brittle rocks. The strength, degree of weathering, resistance to weathering and impact resistance, and the degree of water seepage are different with different rock properties.

(3) Geological structure

The attitude of rock strata is also of great significance to the stability of slopes. If the dip direction of the rock strata is opposite to that of the slope, its stability is greater than that of the slope whose dip direction if along with the rock strata. The degree of stability of the rock strata leaning along the slope also depends on the inclination angle and the degree of fragmentation.

All tectonic actions, normal faults, reverse faults, and overburden faults have an adverse effect on the stability of the slopes. The magnitude of its impact is determined by the properties, size, shape and location of the structural damage. In order to stabilize a single rock slope, iron chain locks or iron clamps can be used to improve the stability of the rock. Fig. 7-22 is a schematic diagram of the relationship between the joint and the collapse process.

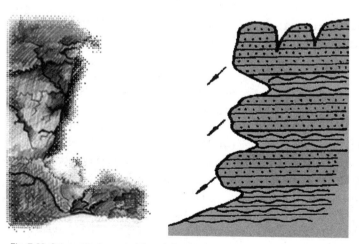
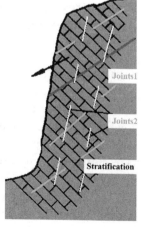

Fig. 7-22 Schematic diagram of the relationship between joint and collapse

(4) Weathering

In particular, differential weathering often forms cliffs-dangerous rocks that cause instability and collapse.

(5) The role of climate and water

Climate change can accelerate weathering, and water infiltration can reduce strength of rock and soil, water erosion of slope toe, all of those can cause collapse.

(6) Human activities

Human engineering construction, such as railways, highways and other transportation projects, may form more excavated slopes, which can induce the redistribution of slope stress and form new cracks (such as unloading cracks), thus promoting the occurrence of collapse.

(7) Other inducing factors

Other inducing factors include earthquakes, heavy rains, floods, etc. Earthquake can trigger a large collapse and landslide geological disasters. Site selection for resettlement and reconstruction planning of roads, railways, schools in the earthquake-stricken areas involve a large number of potential collapse threats. The treatment of collapsed rocks is the priority of post-disaster reconstruction and geological disaster management.

The above (1)~(3) are internal conditions or dominant factors; (4)~(7) are external conditions or inducing factors.

7.3.4 Prevention Measures and Treatments of Dangerous Rock and Collapse

(1) General Principles

Because collapse usually occurs suddenly and violently, it is difficult to manage. The treatment of large or huge collapses is especially very complicated, so people usually take measure, mainly preventive steps, to manage only small collapses:

1) When selecting the construction sites, carefully analyze the possibility of collapse and its scale according to the specific conditions of the slope. The areas where large and medium-sized collapses are likely to occur should be avoided as far as possible. If it is difficult to avoid, preventive works should be taken.

2) In the design and construction, avoid the use of unreasonably high and steep slopes, and avoid large-scale excavation and cutting to maintain the balance and stability of the mountain. In areas where the rock mass is loose or the structure is broken, it is not suitable to use large blasting construction which may result in collapse due to engineering errors.

(2) Prevention and control methods

Measures to prevent dangerous rock collapses include cutting slopes, removing dangerous rocks, cementing rock fissures, guiding surface water flow to avoid rapid changes in rock strength, improving slope stability. For the prevention and control of large-scale and more complex dangerous rock mass, a combination of multiple prevention and control methods should be generally adopted.

1) Blasting or wedging. Slow down the slope of the cliff and remove the rock that is easy to fall.

2) Block the cracks or grouting into the cracks.

3) Adjust the direction of surface water flow. Construct intercepting ditches above the collapsed area to prevent water from flowing into the cracks.

4) To prevent weathering, pave the hillsides and slopes or spray grouting on the slopes.

5) Construct shield or open tunnels to protect the security of roads. As shown in Fig. 7-23.

6) Build retaining walls and enclosures (wood, stone, barbed wire) to block falling rocks, and promptly remove the accumulations in the enclosure buildings.

7) Build retaining walls at the outcropping places of weak rocks to support the quality of the upper rocks (this measure is usually used when the construction of railway subgrade

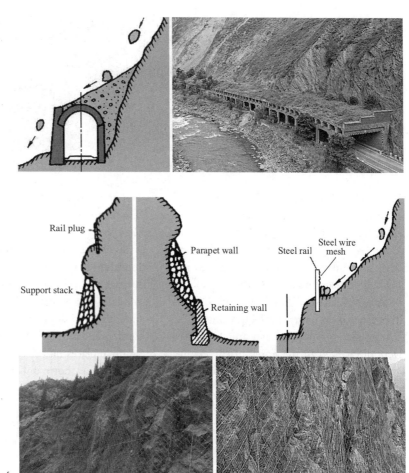

Fig. 7-23 Schematic diagram of the shield

Fig. 7-24 Protections of dangerous rocks

requires deep cutting excavation).

8) Anchorage, support and other reinforcement measures (Fig. 7-24).

9) Flexible protection, including active protection and passive interception (Fig. 7-24).

If the above methods cannot solve the problems, refer to avoidance or tunnel scheme.

7.4 Mudslide

7.4.1 Basic Conceptions

Mudslide refers to a special torrent that is induced by precipitation (i.e., heavy rain in mountainous areas or melting ice and snow, etc.) and is formed on valleys or mountain slopes with a large number of solid materials such as sand, lump stones, and huge particles. This kind of torrent often rushes down from the upper reaches of the valley at a huge speed. It is a kind of geological disaster that can fiercely and quickly destruct downstream buildings and people. The solid debris content in the mudslide is roughly between 20% and 80%.

The geographical distribution of mudslide is widespread. According to incomplete statis-

Fig. 7-25 Catastrophic mudslide in Zhouqu, Gansu province, China

tics, mudslide disasters have hit more than 70 countries and regions in the world, mainly in Asia, Europe, and North and South America. China is one of the countries with the most serious mudslide disasters in the world. Fig. 7-25 shows the sudden heavy rain at 22:00 on August 7, 2010 in Zhouqu County, Gannan Tibetan Autonomous Prefecture in China. The mudstides in Luojiayu and Sanyanyu were discharged to the north of the county, and then rushed to the county from north to south, causing houses along the river to collapse. Mudstides blocked the Bailong River and formed a dammed lake. 1407 people were killed and 358 people missing in this huge mountain torrent disaster.

7.4.2 Classification of Mudslide

Debris flows are classified according to the geomorphological characteristics, material composition, and fluid properties.

(1) According to the landform characteristics of the mudslide catchment area

1) Slope mudslide. Slope mudslide generally doesn't occur in a constant region or has obvious grooves, the outline of which is in bowling shape. It generally develops on slopes above 30°, with shallow-buried underlying bedrock and impervious layer. The mudslide source is mainly the shallow surface layer with small activity scale. Slope mudslide can also occur on the edge of floodplain platforms and ice-water platforms in the northwestern region. The occurrence time and space are not easy to identify, and the disaster scale and loss of a single mudslide are small. If multiple simultaneous mudstides flow into valleys, they can also be transformed into large-scale mudstides. The instability of the slope mudslide is mainly induced by groundwater seepage and subsequent heavy rainfall. Strong winds in the process of heavy rain can cause uprooting and dumping of trees and shrubs to form local damage to the slope. It can occur in multiple places on the same slope in a comb-like shape, as shown in Fig. 7-26(a).

2) Valley mudslide. With the drainage basin as the perimeter and restricted by certain valleys, it has three distinct sections of formation area, transition area, and sedimentary area, with a dumbbell-shaped outline. With the groove as the center, the loose deposits in the provenance area are loosely distributed on both sides of the groove and on the river bed. The collapse, landslide and gully erosion are strong, and the scale of activity is large. The occurrence time and space have a certain regularity and can be identified, and the disaster mode and loss range are large. The main reason is the erosion of the loose source by heavy rains and confluent water. The geological structures have obvious controlling effect on the distribution of mudslide, and most of them are distributed in bands or flakes in the same area, as shown in Fig. 7-26(b).

(2) Based on solid material composition (Table 7-4)

(3) Classification based on the fluid properties of mudslide (Table 7-5)

(4) Classification based on the scale of the mudslide (Table 7-6)

(a) Slope mudslide   (b) Valley mudslide

Fig. 7-26 Typical types of mudslides

Mudslide classification based on material composition   Table 7-4

Types	Material composition	Fluid properties	Residual appearance	Mudslide starting slope	Distribution areas
Mud flow type	Most are silt and clay particles, with uniform particle size. 98% of the particles are less than 2.0 mm in size	Non-Newtonian fluid, viscous, with a viscosity greater than 0.15 Pa/s	Heavy mud residue on the surface	Slow	In loess and volcanic areas
Mudstone type	It can contain clay, silt, sand, gravel, pebble, and boulder, with very uneven grain size	Most are non-Newtonian fluids, and a small part are Newtonian fluids, with viscous and non-viscous fluids	Mud remains on the surface	Steep (slope ratio>10%)	In various geological bodies and deposits
Water stone (sand) type	The content of silt and clay particles is very small, and most of them are particles larger than 2.0 mm. The particle size is very uneven (the flow of water and sand is relatively uniform)	Newtonian fluid, non-viscous	Clean surface, with no mud residue	Relatively steeper (slope ratio>5%)	In igneous or carbonate rock areas

Classification based on the fluid properties of mudslide   Table 7-5

Types	Fluid properties	
	Viscous mudslide	Diluted mudslide
Specific gravity /(t/m³)	1.6~2.3	1.3~1.6
Solid material content (kg/m²)	960~2000	300~1300
Viscosity (Pa/s)	≥0.3	<0.3

Continued

Types	Fluid properties	
	Viscous mudslide	**Diluted mudslide**
Material composition	It includes mainly clay and silt, as well as some gravel, block stones, etc., with corresponding soil and easy-to-weather soft rock formations	It includes mainly gravel and sand, with a small amount of cohesive soil, together with corresponding soil and hard rock formations that are not easy to weather
Sediment characteristics	It is tongue-shaped, undulating, and maintains the characteristics of flow structure. The layer of primary sediments is not obvious. There are "mud balls" in between, but the layers of the major sediments are distinct, and they are not easy to dry up after the flood	It is in the shape of a dragon hill or a fan, and is passable after the flood. The layers are not obvious after drying, with sorting properties
Liquid characteristics	Laminar flow. The solid-liquid two-phase materials move without vertical exchange. The slurry is very thick and has a high buoyancy and suspension force. Rocks are suspended and moved, sometimes rolling. Fluid gusts are obvious, with strong directivity, weak steering and obvious curve climbing	Turbulent flow. The solid-liquid two-phase materials move at unequal speeds, with vertical exchange. The rock flow rate is slower than slurry in rolling or jumping. The mud is relatively turbid and its formation is not obvious. But there are flow and dispersion phenomenon, and water and slurry are easy to leak along the way

Classification of mudslide eruption scale   Table 7-6

Scale	Extra-large mudslide	Large mudslide	Medium mudslide	Small mudslide
Total volume of mudslide at one time $V(\times 10^4 \text{ m}^3)$	$V \geqslant 50$	$10 \leqslant V < 50$	$1 \leqslant V < 10$	$V < 1$
Mudslide peak flow Q (m³/s)	$Q \geqslant 200$	$100 \leqslant Q < 200$	$50 \leqslant Q < 100$	$Q < 50$

Note: As long as any one of the threshold values of "the total volume of mudslide at one time accumulation" and "the peak flow of mudslide" reaches the lower limit of the former level, it should be graded as the former level.

7.4.3 Formation Conditions of Mudslide

The formation and development of mudslide is closely related to the geological, topographic and hydrometeorological conditions in the basin, and profoundly affected by human activities. The main factors are abundant loose solid material, steep topography easy for loose solid material accumulation, enough sudden water source such as rainstorm, as well as human activities.

(1) Topography

In terms of terrain, the area where mudslide can be formed is located with high mountains and deep valley, as well as the large vertical slope of the valley bed. And the drainage basin shape is conducive to gathering water and solid materials. In terms of geomorphology, the mudslide can generally be divided into three parts, i.e., formation area, transition area and sedimentary area. The typical zoning of mudslide valley is shown in Fig. 7-27.

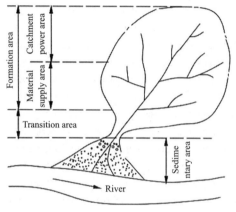

Fig. 7-27 Typical zoning of mudslide

1) Formation area

The formation area is generally located in the upper and middle reaches of the mudslide valley. It can be divided into water catchment power area and solid material supply area. Most of them are small mountain basins surrounded by high mountains. The slopes are steep, the gully bed is cut down, the longitudinal slope is steep, and the water catchment area is relatively large. The area is characterized by broken rock strata, severe weathering, with unstable hillsides, sparse vegetation, serious soil erosion, landslides, and abundant loose deposits. The lithology and denudation strength in the area can directly affect the properties and scale of the mudslide.

2) Transition area

The transition area is generally located in the middle and lower reaches of the river basin, which is mainly valley topography with steep valley walls, narrow river beds, large longitudinal slopes, steep ridges or water drops. After the mudslide enters this area, it has a strong scouring ability to scour the soil and rocks on the valley bed and the valley wall and carrying them away. When the longitudinal slope of the transition area is steep, long and straight, the mudslide flows smoothly and leak straight down, causing great harm. An atypical mudslide valley may not have obvious transition area.

3) Sedimentary area

Most of the sedimentary area is located at the exit of the valley. The terrain is open, and the vertical slope is gentle. The mudslide is diffused so far, with the flow velocity reduced and a large number of solid materials accumulated, forming accumulation fans of different scales. When the accumulation fan is stable and no longer expands, the destructive power of the mudslide slows down and disappears.

The above classification of mudslide areas is only appliable to general mudslide basins. Due to the differences of mudslide, it is often difficult to distinguish clearly. Some transition areas are accompanied by sedimentation. For example, the formation area of slope mudslide is the transition area, and some mudslides are often directly discharged to enter the river and are taken away without obvious accumulation.

(2) Geological conditions

Geological conditions determine the source of loose solid materials. When the water catchment area and the transition area are widely distributed with rock and soil layers of large thickness and soft structure, the easy weathered rocks and soil layers are the main solid materials that provide mudslide source. In addition, mudslides often occur in areas with complex geological structures, well-developed faults and folds, strong tectonic activities, and high seismic intensity. In these areas, broken surface of rock strata, landslides, avalanches, collapse, and other unfavorable geological phenomena can provide a rich source of solid materials for the formation of mudslide.

(3) Hydro-meteorological conditions

Water is not only a component of mudslide,

but also a transport medium for mudslide. After the loose solid material is filled with a large amount of water to reach saturation or supersaturation, the structure is destroyed, the frictional resistance is reduced, and the sliding force is increased, which can result in mudslide. The formation of mudslide is closely related to the sudden surge of large amount of flowing water in a short period of time. The sudden large amount of flowing water comes from heavy rainstorms, short-term intense melting of glaciers and snow, as well as sudden outbursts of glacial lakes, mountain lakes, and reservoirs. Both repeated sudden change of temperature, as well as long-term dryness work to step up the weathering and fragmentation of rocks. Coupled with the softening, suffusion, erosion, and scouring of the rock and soil on the slope by water, the broken materials can increase rapidly, and finally lead to the generation of mudslide.

(4) Human activities

Soil and vegetation directly affect the formation of surface runoff and the particle size distribution of debris transported by mudslide. Indiscriminate logging and reclamation will cause vegetation loss, which can cause hillside losing plant protection, loose soil, and the development of gullies, and greater soil erosion. Those activities could destroy the stability of slope, and form adverse geological phenomena such as landslides and avalanches. As a result, mudslide is easily generated, even retreated mudslide has the possibility of resurrection. Unreasonable excavation of railways, highways, water channels, as well as unreasonable dregs, quarrying, may also result in mudslide.

7.4.4 Features of Mudslide

(1) Generally, the mudslide which is severe with high velocity, and strong paroxysm, contains a lot of loose solid materials such as silt and stones, whose volume content generally exceeds 15%, and weight is generally greater than 13 kN/m^3. The volume content of viscous mudslide solid material can be as high as 80% or more. The flow velocity of mudslide is large, with large variation range from 2.5 to 15 m/s, and such mudslide has strong kinetic energy and damage ability.

(2) The mudslide tends to carry a large number of solid materials, and the flow is sometimes blocked when it encounters a valley turning or obstacle on the flowing path, with some materials accumulated. This would make the valley bed rise rapidly, resulting in ramp turn superelevation, rushing, climbing, to rush over a groove bank or destroy the obstacle. It even cuts the bend and rushes straight out of the new road and rushes downstream. Generally, the more viscous the fluid, the stronger its impact force.

(3) The occurrence of mudslide is periodic. In any area where mudslide can occur, large-scale mudslide does not occur frequently. The occurrence of mudslide has a certain periodicity and can only occur when its conditions permit. After a mudslide occurs, most of the loose materials on the surface of the formation area are washed away. Therefore, it tends to take a long time to gather enough weathered and fragmented materials before the next mudslide. Therefore, the period of mudslide in different regions is different.

(4) The deposits of mudslide are poor in sorting and differentiated greatly in size. Large particles such as crushed stones are poorly rounded and have sharp edges and corners. The surface of the deposits presents ridges and rolling boulders, etc.

These characteristics can be used to identify mudslide, and study the type, frequency, scale, formation history and accumulation speed of mudslide.

7.4.5 Prevention Measures and Treatments of Mudslide

(1) Basic principles of prevention Measures and Treatments

The prevention of mudslide should take advantage of the local treatment, fortification and local materials, and give full play to the effective combination of prevention and control technologies such as draining, blocking, and solidification. The basic principles of protective measures are as follows.

1) Measures to suppress the generation of sediment includes sand dams, valleys, revetments, closures for forests, interception valleys, terraced slopes, etc.

2) Limit the amount of water and the discharged sediment, control the flow path, and prevent erosion and siltation. These measures include sand dams, (i.e., penetrating grid dams), silting yards, diversion dikes, drainage valleys, and riverbed management.

3) To avoid the direct impact of the mudslide, measures should be taken to weaken its energy, and direct the mudslide to the designated area, such as diversion dams, stop dams, ketones, aqueducts, drainage valleys, anti-scouring piers, and protective piles, etc.

(2) Mudslide prevention measures

1) Water and soil conservation.

Generally, water and soil conservation measures are taken in the formation area of mudslide. Soil and water conservation measures includes afforestation by closing hills, planting trees, leveling hillsides, and building terraces. Adjusting surface runoff, building diversion dikes and drainage valleys can make sure that water does not flow along steep slopes and reduce water flow velocity. Strengthening the bank slopes can prevent the erosion and collapse, and reduce the source of solid materials. Soil and water conservation method takes a long time to take effect. Generally, it should be carried out in conjunction with other measures.

2) Retaining measures.

In the transition area of mudslide, the huge energy of the mudslide will be consumed, and the destructive power of the mudslide can be reduced. Specific measures include building various dams, such as the stone masonry dams, grid dams, and overflow earth dams, as shown in Fig. 7-28 and Fig. 7-29.

3) Drainage projects.

Set up drainage measures in the downstream sedimentary area of the mudslide, so that the mudslide can be removed smoothly. It works by improving the flow potential of the mudslide, increasing the flood discharge capacity of bridges and other buildings, so that the mudslide can be discharged smoothly as designation. Drainage projects include flood channels, diversion dikes, rapids, drainage valleys, etc. By the construction of tunnels, open tunnels, and aqueducts (as shown in Fig. 7-30) under the mudslide valley, the mudslide could be drained from above. This is a major engineering measure to ensure railways and highways to pass through the mudslide

Fig. 7-28 Three-dimensional grid dam for preventing mudslide

Fig. 7-29 Sand barrier for preventing mudslide

Fig. 7-30 Three-line cut-and-cover tunnel aqueduct that guides and drains mudstides

area, which is generally used in areas where roadbeds pass through sedimentary areas with large-scale mudslide and frequent serious hazards. For the prevention and control of mudslide, a combination of multiple measures is more effective than a single one.

7.5 Karst and Earth Cave

Karst means all of the phenomena caused by the combined effects of chemical corrosion, mechanical erosion and collapse, during which the soluble rocks are eroded by the surface water and groundwater. It is often accompanied with the formation of grooves, fissures, caves by mechanical action, and the formation of cavern on the surface due to the collapse of its roof.

Earth cave refers to the holes and cavities formed by erosion of surface and underground water, which can lead to ground subsidence and settlement with the extension of cavities.

As a result of the interaction of karst and earth caves, a series of geological problems that are unfavorable to the project can be produced, such as the formation of cavities in rock and soil, the destruction of rock structure, sudden collapse of the ground surface, changes in the circulation of groundwater, etc. These phenomena seriously affect the safety of engineering construction.

7.5.1 Karst

(1) Formation and development conditions of karst

There are many factors for the formation and development of karst, while water-soluble and permeable rocks are necessary, as well as flowing and erosive water.

First of all, the rock mass is soluble. According to the solubility of rocks, the rocks that can cause karst can be divided into three groups, i.e., carbonate rocks, such as limestone, dolomite and marl; sulfate rocks, such as gypsum and anhydrite, as well as halogen rocks, such as rock salt.

The rock mass permeability includes two aspects. One is the water permeability of the soluble rock, which means that there must be voids or fissures in the rock for water flow. The other is the structural fissures and bedding fissures, which have the greatest impact on the development of karst.

Rocks contain corrosive water. Natural water has the ability to dissolve as it contains a certain amount of corrosive CO_2. When the water containing free CO_2 interacts with the calcium carbonate ($CaCO_3$) of the surrounding rock, the calcium carbonate is dissolved, and its chemical effects are as follows.

$$CaCO_3 + CO_2 + H_2O = Ca^{2+} + 2HCO_3^-$$

This part of the carbonic acid consumed to react with calcium carbonate is called corrosive CO_2. It is meaningful to determine the corrosive CO_2 in the water, because the more corrosive CO_2 in the water, the greater the water's ability to dissolve. However, the content of corrosive CO_2 in water varies with the degree of water activity. For this reason, the following paragraph will talk about the activity of water in the rock mass.

The movement of water in soluble rock mass is the main cause of karst. Such movement mainly refers to water flowing in

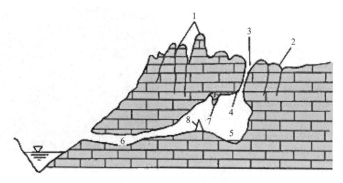

1—Stone Forest; 2—Grooves; 3—Funnel; 4—Swallow hole; 5—Karst cave;
6—Underground River; 7—Stalactite; 8—Stalagmite

Fig. 7-31 Schematic diagram of groundwater karst profile section

the rock mass, and constant alternating of surface water and groundwater. As a result, the water flow has the ability to dissolve the surrounding rock on the one hand, and scour the surrounding rock on the other hand.

The dissolution and erosion of the surrounding rock by groundwater in karst areas occur simultaneously. However, in some fissures or small caves, the dissolution is dominant. In some large underground rivers, the scouring ability of groundwater is very strong, and at this time, the dissolving ability takes a secondary position.

(2) The features and laws of karst development

Features of karst is the geological manifestation in the process of dissolution of soluble rock. It can be divided into surface karst and underground karst. Surface karst includes grooves, stone buds, funnels, karst depressions, sloping valleys, and karst plains. Underground karst includes swallow holes, karst caves, underground rivers and natural bridges (Fig. 7-31).

1) Karst grooves is a tiny topographic form, which is formed on the surface of the rock surface by the dissolution and erosion of surface water, as shown in Fig. 7-32. The grooves cut the surface into jagged, undulating shapes. This kind of landform is called karst grooves wilderness. At this time, the distance between grooves is generally 2 to 3 meters. When the grooves continue to develop, the grooves will connect with each other, with some stalagmite-like rock pillars remaining on the surface. This kind of rock pillar is called stone bud, generally 1~2 m high, and arranged regularly along the joints.

Fig. 7-32 Karst grooves

2) Funnel is a funnel-shaped karst formed on the surface by the dissolution and erosion of surface water, as shown in Fig. 7-33. The size of the funnel varies, with the diameter near the surface as large as hundreds of meters, and the depth of the funnel a few meters. Funnels are often distributed in groups along a certain direction, often along the direction of the structural fracture zones.

3) Karst depression is formed by the continuous expansion and connection of many funnels. It is circular or elliptical on the plane, with a diameter ranging from hundreds of meters to several meters. There are often dissolution mounds, peak clusters, and peak forests around the karst depressions with funnels and swallow holes at the bottom.

4) Slope valley is a large closed karst depression with an area ranging from a few square kilometers to hundreds of square kilometers. The slope valley tends to develop into karst plain. There are often lakes, swamps and wetlands in slope valleys or karst plains whose bottom is often covered by residual alluvial layer or river alluvial layer.

5) Swallow holes are passages from the surface to the deep underground, and most of their lower part is connected with karst caves or underground rivers. It is formed by dissolution, erosion, expansion of rock fissures by water flowing, or collapse of rock fissures. It often appears at the bottom of funnels, grooves, dissolution depressions and slope valleys, or on the edges of river beds. Swallow holes often appear in beaded arrangement.

6) Karst caves are karst forms that develop near horizontally and formed by the long-term groundwater and dissolution, erosion and collapse. The original caves are channels for karst water in the early stage. Therefore, such caves can have various extension and morphology, and there are often branch caves, stalactites, stalagmites, and stone pillars in the caves, as shown in Fig. 7-34. These karst sediments are formed because the dripping water in the cave is bicarbonate water, which

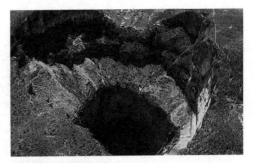

Fig. 7-33 Karst funnel

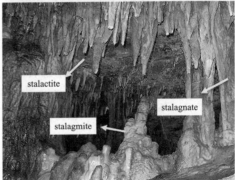

Fig. 7-34 Karst caves

releases CO_2 due to environmental changes to precipitate calcium carbonate.

7) The underground river is the main channel for the collection and discharge of underground karst water, as shown in Fig. 7-35. Some underground rivers are often connected to the grooves, funnels and swallow holes on the ground. The water source of the underground rivers often flows into the underground rivers through the karst grooves and funnels on the ground. Therefore, the development and extension of underground rivers can be roughly judged based on the

Fig. 7-35 The underground river entrance of Tenglong Cave, Lichuan, Hubei, China

Fig. 7-36 A natural bridge in Wulong County, Chongqing, China

distribution of these surface karst forms.

8) Natural bridge. The natural bridge is a stone bridge that spans a river formed by the collapse of a karst cave or collapse of underground river tunnels with part of cave roof remained. Fig. 7-36 is a photo of natural bridge in Wulong County, Chongqing, China.

(3) Forming conditions of karst

The distribution of karst decreases with depth and is controlled by the local karst erosion base level. The development of karst is closely related to the development of fractures and water circulation, and the development of fractures usually decreases with depth. The surface water infiltrates to add to groundwater, and groundwater moves from the water ridge to the surface valley, which will inevitably promote the formation of caves and holes.

The distribution of karst is controlled by lithology and geological structure. Karst will not develop in non-soluble rocks, and its development in weakly soluble rocks will be impeded. Karsts are very developed in pure limestone. The destroy of soluble rocks can promote the development of Karsts. The range of karst development can be classified according to the different solubility of rocks and extent of structural damage in a certain region.

The distribution of karst on the vertical section is often stratified. The crust is often in an intermittent rising and descending phase. In this way, the base level of karst erosion will change accordingly. In order to adapt to the base level, groundwater will practice vertical dissolution, and a vertical channel will be created. The vertical zoning of karst is mainly manifested in Fig. 7-37.

1) Vertical cycling zone. This zone is below the surface and above the water table.

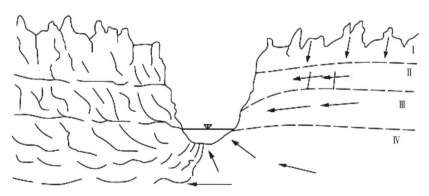

I— Vertical cycling belt; II— Seasonal cycle zone; III— Horizontal cycling zone; IV—Deep cycling zone

Fig. 7-37 Vertical zoning of Karst water

2) Seasonal cycling zone. This zone is located between the lowest and highest groundwater level, affected by seasons.

3) Horizontal cycling zone. This area is located below the lowest groundwater level and filled with water all the year round. The groundwater in this zone flows horizontally or drains into the river valley.

4) Deep cycling zone. The flow direction of groundwater in this zone depends on the geological structure and deep cycling water.

7.5.2 Earth Cave

The earth cave is in a covered karst developing area. Due to its specific hydrogeological conditions, the soil above the bedrock surface tends to migrate to form caves in the soil and collapsed deposits in the caves, resulting in the ultimate ground deformation and destruction. Earth cave is a special form of karst. Because of its rapid development and dense distribution, its impact on the project is far greater than that of rock caves.

(1) Formation conditions of earth cave

Soil, karst and water are the three necessary conditions for the formation of earth caves. The further development of earth caves can lead to ground subsidence.

1) Relationship between earth cave, soil quality and soil layer thickness

The earth cave is mostly located in clay soils, and rare in sand and gravel soils. In cohesive soils, the formation of earth cave depends on conditions such as the composition of clay particles, cohesion, and water properties. The soil layer with fine particles, high viscosity, good cementation, and good hydro-physical properties, provides favorable conditions to form earth caves; otherwise, earth caves are hard to develop.

2) Relationship between earth cave and Karsts

Earth cave are the product of karstification, and their distribution is controlled by factors such as lithology, karst water and geological structure. In all karst areas that meets the requirement for earth cave creation, earth caves are available.

3) Relationship between earth cave and groundwater

Water is the external cause and driving force for the formation of earth cave. Therefore, the distribution law of earth cave must obey the law of interaction between soil and water. The earth caves formed by groundwater are mostly within the range of groundwater changes, and most of them are distributed between the high water table and the low water table.

(2) Types of earth caves

According to the characteristics of earth caves and the driving force of water, the formation of earth caves in China can be divided into mechanical erosion ones and chemical erosion ones.

1) Earth caves caused by mechanical erosion

There are three main factors for the formation of earth cave formed by mechanical erosion of surface water seepage.

① The proper properties of the soil layer are the basis for the development of earth cave. The sandy silt layer with gravel could provide good conditions for surface water to penetrate downwards to induce corrosion.

② There is a channel under soil layer is for water flow and soil particle discharge, resulting in the intense development of earth cave.

③ Surface water flow can directly infiltrate the soil layer. The fissures in the soil layer are caused by long-term drought, which causes shrinkage fissures on the surface. This would become a channel for rains to constantly erode downward. Under the action of hydraulic force, the collapsed soil particles are taken away to form earth cave, which will continue to develop until the roof is destroyed and the ground collapses.

2) Earth caves caused by chemical erosion

Such earth cave is hydraulically connected to karst water. They are distributed in the

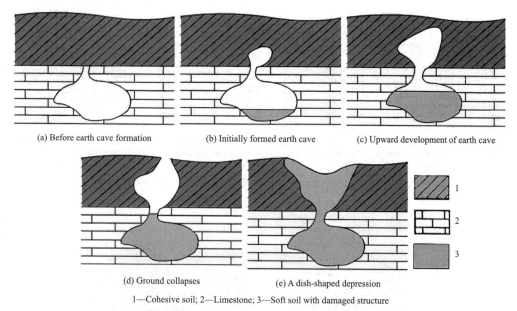

Fig. 7-38 Development process of earth cave caused by chemical erosion

1—Cohesive soil; 2—Limestone; 3—Soft soil with damaged structure

contact area between the bedrock surface and the overlying saturated soil layer in karst areas. The formation of such earth cave is due to a saturated soft plastic to semi-fluid soft soil layer at the contact between the bedrock surface and the overlying soil layer in the karst area. When groundwater moves near the surface of the karst bedrock, the rise and fall of the water can soften the soft soil layer, and erode the soil layer to take away its soil particles. When the earth cave is continuously eroded and washed away, the earth cave gradually expands until the roof cannot bear the upper pressure. The ground gradually sinks or the whole block collapses to form the butterfly-shaped, basin-shaped, grooved and shaft-shaped karst depressions (Fig. 7-38).

7.5.3 Engineering Geological Problems of Karst and Earth Cave

Karst and earth cave have a great influence on the stability and safety of buildings.

(1) The strength of dissolved rocks is greatly reduced. The dissolution of karst water in the soluble rock mass can cause holes and caves. The most common holes are dissolved voids or small holes in the rock mass. Dissolution of the rock mass can make the rock mass have holes and loose structure to reduce rock strength and increase the water permeability.

(2) Karst and earth cave tend to cause the uneven undulation of bedrock surface due to the existence of stone buds and karst grooves. This would cause the unevenness of the foundation and traffic difficulties, as shown in Fig. 7-39.

7.5.4 Prevention Measures and Treatments of Karst and Earth Caves

Before engineering design, the locations of karsts and earth caves should be investigated clearly, and then the corresponding prevention and control measures should be taken. The threatening karst and earth cave areas should be avoided to ensure the construction safety.

(1) Excavation and filling

Dig out weak fillings in karst and earth caves, and backfill with gravel, block stones or concrete, etc. And then tamp them in layers to

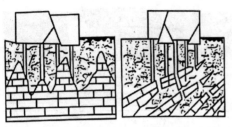  

(a) Uneven foundation settlement caused by the unreliable pile end support due to the uneven horizontal and inclined soluble rock bedrock surface (b) Structural cracking caused by the soil collapse near the Karst cave (c) Structural cracking caused by the shearing—sliding along the inclined karst bedrock surface

Fig. 7-39 Uneven foundation settlement and unreliable support of piles due to the uneven undulation of bedrock surface

 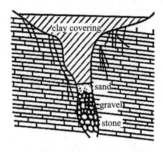

(a) Dry or slurry stone backfilling; (b) Loose rock and soil backfilling

Fig. 7-40 Karst cave backfilling

improve the foundation, as shown in Fig. 7-40.

(2) Crossing cover

When the cave is deep buried or the roof of the cave is unstable, the crossing cover scheme can be adopted. It means crossing the karst caves with beam foundation or arc structure, and covering the karst caves by plate foundation with high stiffness.

(3) Grouting

For some deep buried karst caves with large volume, grout cement mortar or concrete to fill the whole karst caves.

(4) Drainage

The movement of the water in the cave can make the cave sidewall and the cave roof dissolve, scour or be eroded, further resulting in collapse.

(5) Piling

When the earth cave is deep buried, set up pile foundation, such as concrete piles, wooden piles, sand piles or blasting piles.

7.6 Land Subsidence and Ground Collapse

7.6.1 Land Subsidence

(1) Classification and hazards of land subsidence

The broad meaning of land subsidence refers to the geological phenomenon of surface elevation loss with a certain scale and in a certain area, caused by natural forces or human economic activities on the surface of the earth's crust. The subsidence speed and settlement value, as well as the duration and range, are different due to the specific inducing factors or different geological environments.

Present study of land subsidence mainly focuses on the ground elevation of a region decreasing due to compaction deformation caused by artificial exploitation of groundwater, oil, and natural gas. Land subsidence can be divided into the following categories, as shown in Table 7-7 and Table 7-8.

The engineering hazards of land subsidence mainly include two aspects. One is the hazards to the environment, such as tides crossing the embankment and land waters. The other is the hazards to construction projects, such as the sinking of bridge piers, the reduced clearance under bridges, the sinking of docks and warehouses (Fig. 7-41), the changes in the slope of underground pipelines, the relative rise of the feet of deep well pipes and pile foundations, and tilting of buildings, etc.

The land subsidence in Shanghai, Tianjin,

Fig. 7-41 Foundation detachment caused by land subsidence

Xi'an, Taiyuan and other cities in China once severely affected urban planning and economic development, deteriorating the urban geological environment, and causing great losses to the national economy.

(2) Causes of land subsidence

The causes of ground subsidence mainly include three aspects, i.e., over-exploitation of groundwater, under-consolidation of

Classification of land subsidence by dominant factors   Table 7-7

Types	Classification description
Consolidated land subsidence	(1) Land subsidence caused by consolidation of under-consolidated soil, such as natural consolidation of soil; (2) Swabbing ground subsidence caused by the extraction of a large amount of underground liquid and gas resources; (3) Loading ground subsidence caused by compaction of foundation soil due to major construction and water storage projects; (4) Dynamic disturbed land subsidence caused by soil compaction deformation due to ground vibration from large machinery, motor vehicles and blasting; (5) Land subsidence caused by construction and building load due to urban construction
Unconsolidated land subsidence	(1) Land subsidence due to the natural actions, such as tectonic land subsidence, sea-rise land subsidence, seismic land subsidence, volcanic land subsidence, freeze-thaw evaporation land subsidence; (2) Land subsidence caused by the extensive goaf formed by underground mining and the development of underground engineering

Classification based on the scale of land subsidence   Table 7-8

Scale	Giant land subsidence	Large land subsidence	Medium land settlement	Small land subsidence
Subsidence area S (m²)	$S \geqslant 500$	$100 \leqslant S < 500$	$10 \leqslant S < 100$	$S < 10$
Cumulative settlement h (m)	$h \geqslant 1.0$	$0.5 \leqslant h < 0.1$	$0.1 \leqslant h < 0.5$	$h < 0.1$

foundation soil, additional loads of high-rise buildings and dynamic vibration such as traffic loads.

1) Over-exploitation of groundwater

Confined water is often used as a source of industrial and domestic water. In confined aquifers, the extraction of groundwater decreases the confined water table. According to Terzaghi's effective stress principle ($\sigma=\mu+\sigma'$), when water is pumped in the aquifer and the water table drops, the total stress (σ) in the relatively water-resistant clay layer remains approximately unchanged. The void water pressure (μ) borne by the void water decreases, and the effective stress (σ') borne by the solid particles will increase accordingly, resulting in dense soil layers and settlement deformation of the ground surface.

2) Under-consolidation of foundation soil

The transformation of void water pressure to effective stress in the under-consolidated clay of the foundation soil is not as sharp as that of sand layers, so the change of the void ratio in the clay is slow. The compression deformation of cohesive soil also takes a certain amount of time. Therefore, the under-consolidated soil layer will cause ground settlement as the void pressure dissipates.

3) Additional loads

Additional loads due to high-rise buildings and dynamic vibration such as traffic loads will increase the total stress in the soil layer, and the effective stress (σ') borne by the solid particles will also increase, resulting in soil compaction and the ground settlement.

(3) Analysis of ground subsidence

Draw contour maps of land subsidence at different times to analyze the change dynamics of the land subsidence center and its relationship with the groundwater table drop funnel, as well as the relationship between ground rebound and the groundwater table anti-funnel. Analyze the changing laws of land subsidence at different times, at different locations, and under different conditions of groundwater exploitation. Plot special engineering geological zoning maps with land subsidence according to the comprehensive geological conditions of accumulated subsidence and annual subsidence speed.

7.6.2 Ground Collapse

(1) Concept and classification of ground collapse

Ground collapse refers to a dynamic geological phenomenon in which the surface rock and soil mass collapse, ground depressions and potholes are formed under the action of natural or human factors. The ground manifestations are cracks, uneven subsidence and sudden subsidence of the surface rock and soil in a local area.

Ground collapse can be classified according to its dominant factors and scales, as shown in Table 7-9 and Table 7-10.

Classification of ground collapse by dominant factors   Table 7-9

Types	Classification description
Karst ground collapse	Ground collapse due to the imbalance of the roof of the overlying rock and soil layer under the action of natural or man-made factors.
Mined-out ground collapse	Ground collapse caused by underground excavation activities that lead to mined-out area and the rock and soil mass losing support.
Other ground collapse	Ground collapse caused by natural effects (such as water infiltration, water table fluctuation, gravity, etc.); swabbing ground collapse caused by extraction of a large amount of groundwater and gas resources

Classified by the scale of ground collapse   Table 7-10

Scale level	Giant ground collapse	Large surface collapse	Medium surface collapse	Small ground collapse
Diameter of collapse pit D (m)	$D \geqslant 50$	$30 \leqslant D < 50$	$10 \leqslant D < 30$	$D < 10$
Impact range S (km^2)	$S \geqslant 20$	$10 \leqslant S < 20$	$1 \leqslant S < 10$	$S < 1$

(2) Hazards of ground collapse

Ground collapse tends to destroy the urban construction facilities and endanger people's lives and property. For example, densely-populated Tongling City, Anhui Province, China witnessed serious ground collapse. The long-term drainage of Tongguanshan Copper Mine has created conditions for the subsurface erosion and hollowing of karst caves in this area, which leads to many ground collapses finally. Road and railway traffic was interrupted, and various pipelines were broken and twisted, causing serious economic losses.

Ground collapse tends to affect the development and utilization of natural resources, including surface water, karst groundwater and mineral resources. For example, in 1978, the water inrush from the roadway caused the ground collapse of the Kejiagou River Valley, with more than 70 swallow holes produced. The river water was cut off due to a large amount of leakage into the ground. Buildings within 4,000 meters on the bank was destroyed, and railway and high-voltage transmission lines for mines were damaged, causing nearly one million yuan in economic losses.

7.6.3 Prevention Measures and Treatments of Ground Subsidence and Collapse

Once land subsidence occurs, it is difficult to recover. Therefore, for urban areas where land subsidence has occurred, on the one hand, treatment measures should be taken according to the geographical environment and the degree of disasters, to reduce or eliminate the hazards. On the other hand, timely and proactive measures should be taken to control the development of land subsidence, on the basis of finding out the causing factors of land subsidence.

(1) Treatments of land subsidence

For areas where land subsidence has occurred, the following control and treatment plans can be adopted according to engineering geological and hydrogeological conditions.

1) Reduce the amount of groundwater extraction. When the land subsidence develops violently, the exploitation of groundwater should be temporarily stopped.

2) Recharge groundwater artificially. Water quality of recharged water should be controlled to prevent groundwater from being polluted, and reasonable mining and recharge plans should be formulated in accordance with groundwater dynamics and land subsidence laws.

3) Investigate the levels of groundwater mining. Reasonably, and properly mine the deep groundwater or karst fissure water.

(2) Treatments of ground collapse

The technical measures to prevent ground collapse include water control measures and engineering measures. As for water control measures, reasonable control of the groundwater table is important in groundwater mining areas. In mine drainage, local grouting treatments may be applied to areas where ground collapse may occur. When loose soil layers are drained, the pumping of wells must be controlled. Engineering measures include backfilling, blocking and reinforcement. Backfilling means using clay or slag to fill the pit and tamping. For the ground collapse caused by drainage, curtain grouting or water cut-off wall can be used to block the underground water flow. For the collapse caused by surface water flow, dam and cofferdam can be constructed to isolate. For cave collapse, concrete plug can be used to plug cave entrance. When the building foundation collapses, it can be reinforced by methods such as pile support and foundation reinforcement.

Quiz

1. What is the correlation and difference between the earthquake magnitude and intensity? What are the types of seismic waves and their propagation and destruction characteristics?

2. What are the types and characteristics of secondary geological disasters induced by earthquakes and their impact on the site selection for the post-earthquake reconstruction?

3. What are the destructive effects of earthquakes?

4. What is a landslide? How to classify landslide? Try to analyze the internal conditions and external incentives of landslide respectively.

5. How to identify landslides in the field? What are the preventive measures for landslides?

6. What is the role of crack investigation in the stability of landslides?

7. Describe the classification of landslides according to mechanical conditions and their occurrence conditions and deformation characteristics.

8. What is the dangerous rock? What is the collapse?

9. Try to analyze the conditions and developmental factors for the collapse. Try to analyze what are the collapse prevention measures with specific engineering cases.

10. What is mudslide? What are the conditions for its formation? What are the commonly used prevention and treatment measures?

11. What are the movement characteristics of mudslide?

12. What is karst? What are the formation and development conditions of karst? Briefly describe the main engineering geological problems of karst site foundation. What are the treatment measures of karst construction sites?

13. What is land subsidence? Analyze the main causes of ground surface subsidence and the main preventive measures.

Chapter 8
Engineering Geology Investigation

8.1 Grades and Stages of Engineering Geological Investigation

8.1.1 Investigation Grades

In the current Code for Investigation of *Geotechnical Engineering* (GB 50021—2001) (2009 edition), the grade of engineering investigation is mainly determined according to the engineering importance grade, the site grade of the engineering location and the foundation grade. On this basis, people carry out the engineering investigation. The division of engineering importance grade, site grade and foundation grade are shown in Table 8-1~Table 8-3.

The grade of engineering investigation

Classification of project importance grades   Table 8-1

Engineering importance grades	Project types	Damage consequence
Grade I	Important project	Very serious
Grade II	General engineering	Serious
Grade III	Secondary project	Not serious

Note: Relevant regulations and descriptions of engineering scale and damage consequences can be found in relevant design codes for different project types.

Classification of site grades   Table 8-2

Site grades	Division standards (divided according to the complexity of the site)
Grade I (Complex Site)	Compliance with one of the following conditions: 1. Dangerous areas for earthquake resistant buildings; 2. Bad geological action develops strongly; 3. Geological environment has been or may be strongly damaged; 4. The topography is complex; 5. Complex hydro-geological conditions including multilayer groundwater, karst fissure water
Grade II (Medium complex site)	Compliance with one of the following conditions: 1. The area unfavorable to earthquake resistant buildings; 2. Adverse geological processes are generally developed; 3. The geological environment has been or may be generally damaged; 4. Topography and landforms are complex; 5. The foundation lies below the water table
Grade III (Simple site)	Compliance with all the following conditions: 1. The seismic fortification intensity is⩽6 degrees, or the section is favorable for building seismic resistance; 2. Undeveloped unfavorable geological process; 3. The geological environment is basically undamaged; 4. Simple landform; 5. Groundwater has no impact on projects

Note: The dangerous, unfavorable and favorable division of seismic sections of buildings is implemented according to *"Code for seismic design of buildings"*(GB 50011—2010) (2016 edition).

Classification of foundation grades Table 8-3

Foundation grades	Classification standards (according to the complexity of foundation)
Grade I (Complex Foundation)	Compliance with one of the following conditions: 1. There are many kinds of rock and soil, which are very uneven and changeable in properties, which need special treatment; 2. Special soil with serious collapsibility, expansion, salinity and pollution and other rock and soil requiring special treatment
Grade II (Medium complex foundation)	Compliance with one of the following conditions: 1. There are many kinds of rock and soil, which are uneven and changeable in properties; 2. Special rock and soil other than those specified for Grade I foundation
Grade III (Simple foundation)	The type of rock and soil is single and uniform, the property changes little, and there is no special rock and soil.

Classification of engineering investigation Table 8-4

Investigation grades	Division criteria
Grade A	One or more of the engineering importance grades, site grades and foundation grades are Grade I.
Grade B	Investigation items other than Grade A and Grade C
Grade C	The project importance grade, site grade and foundation grade are all Grade III

Note: For the Grade I project constructed on rock foundation, when the site complexity and foundation complexity are both Grade III, the geotechnical engineering investigation level can be set as Grade B.

shall be comprehensively analyzed and determined according to the importance grade of construction engineering, the grade of construction site and the grade of building foundation shown in Table 8-4.

8.1.2 Investigation Stages

In order to provide the engineering geological data required in each design stage, the investigation work is divided into three stages: feasibility study investigation (site selection investigation), preliminary investigation, and detailed investigation. For important building foundations with complex engineering geological conditions or special construction requirements, pre-feasibility and construction investigations should be carried out; for sites with simple geological conditions and a small building area, or areas with construction experience, simplify the investigation phase appropriately.

8.2 Methods of Engineering Geological Investigation

The methods or technical means of engineering geological investigation mainly include the following: engineering geological mapping; exploration and sampling; in-situ test and laboratory test; field inspection and monitoring.

8.2.1 Engineering Geological Mapping

(1) Basic concepts

Engineering geological mapping is the preparation of engineering geological maps by using different colors, symbols, according to the precision requirements of plotting on the topographic map of a certain scale. Before mapping, it's necessary to use geology and engineering geological theory to observe and describe various geological phenomena related to engineering construction, and to preliminarily identify the engineering geological conditions of the proposed site or construction site. According to different research contents, engineering geological investigating and mapping can be divided into two kinds: comprehensive investigating and mapping, and specialized investigating and mapping.

(2) Scale of mapping

The scope of engineering geological mapping and investigation should include the site and its nearby sections. The scale of mapping is as follows:

Feasibility study investigation: 1 : 5000~1 : 50000;

Preliminary investigation: 1 : 2000~1 : 10000;

Detailed investigation: 1 : 500~1 : 2000.

(3) Layout, density and positioning of geological observation points

The layout, density and location of geological observation points shall meet the following requirements:

1) There should be geological observation points on the lines of geological structure, stratigraphic contact, and lithologic boundary. Standard horizon and each geological unit body should have geological observation points.

2) The density of geological observation points shall be representative and determined according to the geomorphology, geological conditions, mapping scale and engineering characteristics of the site.

3) Geological observation points should make full use of natural and artificial outcrops; When the outcrop is small, a certain number of pits or grooves should be arranged.

4) The location of geological observation points should be selected according to the accuracy requirements of the appropriate method. Special geological observation points such as geological structure line, stratigraphic contact line, lithologic boundary line, weak interlayer, groundwater outstrip and adverse geological phenomena should be located with instruments.

(4) Items of engineering geological investigating and mapping

Engineering geological mapping and investigation mainly includes the following items:

1) Identify the topographic and geomorphic features and their relationship with strata, structures and undesirable geological phenomena, and divide the geomorphic units;

2) Identify the property, origin, age, thickness and distribution of rock and soil, and the weathering degree of rock strata should be identified, and distinguish the newly sedimentary soil and various special soils;

3) Identify the attitude and structure types of rock strata, the attitude and properties of weak structural plane, including the location, type, attitude, fault spacing, width of broken zone and filling and cementation, the contact surface of rock and soil layer and the characteristics of weak interlayer, the traces and characteristics of quaternary tectonic activity and the relationship with seismic activity;

4) Find out the type of groundwater, supply source, discharge conditions, location of wells and springs, lithologic characteristics of aquifer, burial depth, water table change, pollution and its relationship with surface water;

5) Collect meteorological, hydrological, vegetation, soil and other data of the maximum

freezing depth, investigate the highest flood level and its occurrence time and inundation range;

6) Find out the formation, distribution, form, scale, development degree of karst, soil cave, landslide, mudslide, collapse, gully, ground subsidence, fault, earthquake damage, ground crack and shore scour of adverse geological phenomena and their influence on engineering construction;

7) Investigate the impact of human activities on construction site stability including drainage and reservoir induced earthquake;

8) Get to know building deformation and engineering experience.

(5) Engineering geological mapping methods

1) Photo mapping means using ground photography or aerial (satellite) photography to describe the identified stratigraphic lithology, geological structure and landform, water system and adverse geological phenomena according to the interpretation marks indoors and known regional geological data. Then conduct field investigation for a number of points and routes selected to be investigated in the photo for proofreading and correction, and draw the base map. Finally, transform all the information into engineering geological map.

2) Field investigating and mapping is the field work method, which is subdivided into the following three methods: the route method, the point distribution method, tracing method.

3) Remote sensing technology measures are to record the properties and characteristics of long-distance targets through highly sensitive instruments and equipment. Based on the basic electromagnetic wave theory, obtain the necessary geological and geomorphic information by observing the terrain near the surface and the electromagnetic spectrum emitted (or reflected) by surface objects. The information is a basis for solving relevant problems.

There are two recording methods for remote sensing data: non-imaging and the imaging.

(6) Data and results of engineering investigating and mapping

1) Check field data

① Check whether the items described in various field records are complete;

② Check in detail whether the strata, lithology, structure, landform and geological origin boundaries divided by various original maps are consistent with the actual situation in the field, and whether the boundaries in different maps are consistent with each other;

③ Check whether various geological phenomena filled in the field are correct;

④ Check whether the collected data are consistent with the investigating and mapping data. In case of any contradiction, analyze the causes;

⑤ Sort out and check various specimens collected in the field.

2) Documents after sorting: the achievement data of engineering geological mapping and investigation generally include the actual material map, comprehensive engineering geological map or engineering geological zoning map, comprehensive geological histogram, engineering geological section map and various sketch maps, photos and text descriptions.

8.2.2 Exploration and Sampling

Engineering geological exploration methods mainly include drilling, well exploration, trench exploration, cave exploration and geophysical exploration, as shown in Fig. 8-1~Fig. 8-4.

(1) Engineering Drilling

According to the methods of drilling, it can be divided into four types: percussion drilling, rotary drilling, vibration drilling and flushing drilling. The above four methods have their own characteristics, and are suitable for different investigation and rock and soil layers of various properties, as shown in Table 8-5.

Fig. 8-1 Adit exploration

Fig. 8-2 Trenching exploration

Fig. 8-3 Well exploration

Fig. 8-4 Geo-physical exploration

(2) Take soil samples

According to the sampling methods and test purposes, the disturbance degree of soil samples is divided into four grades. The names of each grade and the test items are shown in Tables 8-7.

In order to meet the requirements of different grades of soil samples, use required sampling methods and tools. According to *"Code for Investigation of Geotechnical Engineering"* (GB 50021—2001) (2009 edition), the application scope and sampling quality of different samplers are shown in Table 8-8.

When taking grade-Ⅰ and grade-Ⅱ soil samples in drilling, the following requirements should be met:

Application scopes of different drilling methods   Table 8-5

Method of drilling		Drilled formation					Investigation requirements	
		Clay soil	Silt	Sand	Gravel soil	Rock	Undisturbed soil sample	Disturbed soil sample
Rotary	Auger drilling	√	○	○	×	×	√	√
	Coreless drilling	√	√	√	○	√	×	×
	Core drilling	√	√	√	○	√	√	√

Continued

Method of drilling		Drilled formation					Investigation requirements	
		Clay soil	Silt	Sand	Gravel soil	Rock	Undisturbed soil sample	Disturbed soil sample
Percussion	Percussion drilling			√	√	×	×	×
	Hammer drilling	√	√	√	○	○	√	√
Vibration drilling		√	√	√	○	○	○	√
Flush drilling		○	√	√	×	×	×	×

Note: 1. √ Applicable; ○Partially applicable; ×Not applicable;

2. In drilling, there are mainly three tasks for engineering geologists: first, compile the design program as the basis for drilling; second, observe and catalog the core during the drilling process; third, organize the data internally after the completion of the drilling. For details, as shown in Table 8-6.

Steps and contents of drilling work Table 8-6

Steps	Specific content of each step
Drilling design program preparation	① The terrain and geology near the borehole; ② The purpose of the borehole and the problems that should be paid attention to during drilling; ③ The type, depth, and structure of the borehole, the drilling method, the drilling speed and the way of fixing the wall, etc.; ④ Engineering geology requirements, including core sampling rate, sampling, in-hole test, observation and water-stopping, etc.; ⑤ Prepare after-drilling guidance to decide whether reserve the drill hole for long-term observation or seal it. Engineering geologists should compile a drilling geological profile in the task program so that the drillers can grasp the location of some important horizons, strengthen the drilling management, and determine the drilling type, hole depth and hole structure
Observation and cataloging of boreholes	① Core observation, description and cataloging work: during the drilling process, each time the footage is generally 0.5~0.8 m (maximum 2 m), the core needs to be taken. The core rate of the whole hole is not less than 80%, and the minimum is not less than 60%. The core should be carefully observed and identified to determine the name and the physical properties of the rock and soil. Arrange and number the cores in order, and record the statistics, including statistics of core acquisition rate and rock quality indicators; ② Hydrogeological observation: Observation records should be made for the groundwater level and dynamics in the borehole, including the water level elevation, thickness of the aquifer, groundwater temperature, water quality, and flushing fluid consumption during drilling; ③ Drilling record and description: During the drilling process, make records of situations including the drilling tool sinking, strong vibration, hole wall collapse, water gushing and other phenomena
Drilling data collation	① Prepare borehole histogram; ② Fill in operation and hydrogeological log; ③ Make core sketch. The above three documents are graphical representation essentially of the aforementioned work, and the final drilling results, which must be carefully sorted and compiled for archiving and review

Classification of quality grades of soil samples　　　　　Table 8-7

Grade	Disturbance	Test items
I	Undisturbed	Soil naming, water content, density, strength test and consolidation test
II	Slight disturbance	Soil naming, water content, density
III	Significant disturbance	Soil naming and water content
IV	Completely disturbed	Soil naming

Note: 1. Non-disturbance means that although the in-situ stress state has changed, the structure, density and water content of soil change little, which can meet the requirements of laboratory tests.

2. If there are no conditions for grade I soil sampling, grade II soil samples can be taken as an alternative if engineering technology permits after identifying the disturbance degree of soil samples to determine the suitability of the test. The test results shall be used together with regional experience.

1) Slurry wall protection should be used in soft soil and sandy soil. If the casing is applicable, the water table in the pipe shall be kept equal to or higher than the groundwater table, and the sampling position shall be less than 2 times the hole diameter at the bottom of the casing.

2) When drilling by methods like flushing, strike, and vibration, turn to use rotary drilling over 1 meter above the expected sampling position.

3) Boreholes should be carefully cleared and disturbed soil should be removed before the soil sampler is laid down. The thickness of residual floating soil at the bottom of the hole should not be greater than the length of the waste soil section of the soil sampler.

4) Rapid static continuous pressing method is recommended for soil samples.

(3) Engineering geological well, trench and hole exploration

The pit and trench exploration projects commonly used in engineering geological exploration are: trench, test pit, shallow well, shaft and peace hole, as shown in Fig. 8-5. The first three are light pit and trench exploration projects, and the last two are heavy ones. The characteristics and applicable conditions of various pit and trench exploration projects are listed in Table 8-9.

8.2.3 In-situ Testing Technology

Compared with the laboratory test, in-situ test has the following advantages:

① The engineering mechanical properties of undisturbed soil samples that are difficult to obtain, such as saturated sand, silty soil, and flow-plastic silt or muddy soil, can be measured;

② The influence range of rock and soil mass is much larger than that of indoor samples, so such samples are more representative.

③ Complete stratigraphic profile and physical and mechanical indexes can be obtained as many in-situ test methods can be carried out continuously.

④ In-situ testing is fast and economy,

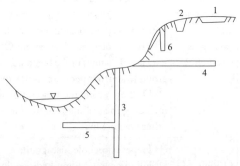

1, 2—trench; 3—shaft; 4, 5—tunnel; 6—shallow well

Fig. 8-5 Schematic diagram of pit types commonly used in engineering geology

Sampling tools or methods required for different grades of soil samples　　　　Table 8-8

Quality grade of soil sample	Sampling tool or method		Applicable soil								Sand			Gravel sand, gravel soil, soft rock
			Clay soil					Silt	Silt	Fine sand	Medium sand	Coarse sand	
			Flow plastic	Soft plastic	Plastic	Hard plastic	Hard						
I	Thin wall soil sampler	Fixed piston	++	++	+	—	—	+	+	—	—	—	—
		Water pressure fixed piston	++	++	+	—	—	+	+	—	—	—	—
		Free piston	—	+	++	—	—	+	+	—	—	—	—
		Exposure	+	+	+	—	—	+	+	—	—	—	—
	Rotary soil extractor	Single Action Triple Tube	—	+	++	++	+	++	++	++	—	—	+
		Double acting triple tube	—	—	—	+	++	—	—	—	—	—	—
	Engraved block soil samples in exploration wells (trenches)		+	++	++	++	++	+	+	+	—	—	—
II	Thin wall soil sampler	Water pressure fixed piston	++	++	+	—	—	+	+	—	—	—	—
		Free piston	+	++	++	—	—	+	+	—	—	—	—
		Exposure	++	++	++	—	—	+	+	—	—	—	—
	Rotary sampler	Single moving triple tube	—	+	++	++	+	++	++	++	—	—	—
		Thick-walled open sampler	—	—	—	+	++	—	—	—	++	++	++
	Thick-walled open sampler		+	++	++	++	++	+	+	+	+	+	—

Continued

Quality grade of soil sample	Sampling tool or method	Applicable soil										
		Clay soil				Silt	Sand					Gravel sand, gravel soil, soft rock
		Flow plastic	Soft plastic	Plastic	Hard plastic	Hard		Silt	Fine sand	Medium sand	Coarse sand	
III	Thick-walled open soil extractor	++	++	++	++	++	++	++	++	++	++	—
	Standard penetrator	++	++	++	++	++	++	++	++	++	++	—
	Thread drill	++	++	++	++	++	+	—	—	—	—	—
	Core bit	++	++	++	++	++	++	+	+	+	+	+
IV	Standard penetrator	++	++	++	++	++	++	++	++	++	++	—
	Thread drill	++	++	++	++	++	+	—	—	—	—	—
	Core bit	++	++	++	++	++	++	++	++	++	++	++

Note: 1. ++ applicable, + partial use, — not applicable.
2. Additional measures shall be taken to prevent the loss of sand samples.
3. Beam-type earth samplers shall be used instead of thin-walled ones if the worker has experience.

Types of pits and trench exploration in engineering geological exploration Table 8-9

Type	Characteristics	Applicable conditions
Trench	A long strip groove with a depth of less than 3~5 m in the vertical rock formation or structure line on the surface	Strip the overlying soil on the ground, expose the bedrock, and classify the lithology of the strata; explore the residual slope accumulation layer; study the fault fracture zone; understand the geological conditions at the dam joints
Trial pit	Straight from the surface, round or square with depth less than 3~5 m pit	Local stripping of surface soil, uncover bedrock, determine the lithology of the formation. Load test, seepage test, undisturbed soil samples
Shallow shaft	A round or square well with a depth of 5~15 m straight down from the surface	Determine the lithology and thickness of cover and weathering layer. Load tests, undisturbed soil samples
Shaft (Inclined shaft)	Same shape as shallow well, but deeper than 15 m, sometimes supporting is needed	Applied in gentle slopes, floodplains, terraces and other places where the strata are relatively gentle to understand the thickness and properties of overburden, the thickness and lithology of weathering crust, the distribution of weak interlayers, fault fracture zones and karst development, landslide structure and sliding surface, etc.
Flat hole	Horizontal tunnels with exits on the ground and greater depth	Applied in a bedrock slope with steep terrain to investigate the geological structure of the slope. It helps on the identification of stratigraphic lithology, weak interlayer, fracture zone and weathered rock strata in the valley area. It can also be used for sampling and in situ rock mechanics test and ground stress measurement

which can greatly shorten the investigation period.

Although having the above advantages, it has certain limitations. For example, various in-situ tests have strict applicable conditions. If improperly used, the effect will be affected, and even the wrong results will be obtained. Some large-scale tests are time-consuming and costly, and should not be carried out in large quantities. The relationship between the parameters obtained from many in-situ tests and the engineering properties of rock and soil is based on a large number of statistical empirical relationships. Therefore, in-situ tests of rock and soil should be carried out in conjunction with laboratory tests.

There are many methods for in-situ testing, which can be divided into three categories:

① Geotechnical mechanical properties test. For example, load test, static penetration test, cone dynamic penetration test, standard penetration test, vane shear test, side pressure test, flat shovel side expansion test, field shear test, rock and soil in-situ stress test, acoustic test, point load test, as shown in Fig. 8-6~Fig. 8-9.

② Hydro-geological test: For example, drilling pumping test, water pressure test, seepage test.

③ Test of improving geotechnical properties. For example, grouting test, pile bearing capacity test.

This section focuses on geotechnical mechanical properties test.

(1) Field test on mechanical properties of soil

1) According to the shape of the bearing

Fig. 8-6 Plate loading test

Fig. 8-7 Vane shear test

Fig. 8-8 Standard penetration test

Fig. 8-9 Cone penetration test

plate, the static load test can be divided into plate load test and spiral plate load test.

① The device of load test is shown in Fig. 8-10. The device of load test is composed of pressure plate, loading device and settlement observation devices. The pressure plate is generally square or circular plates. The loading devices include pressure source, load bench or reaction frame. The two loading methods are heavy loading or oil pressure jack pressure loading. The settlement observation devices include percentage meter, settlement sensor

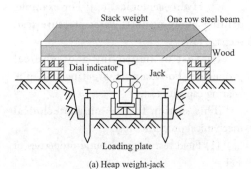

(a) Heap weight-jack

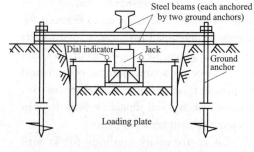

(b) Ground anchor-jack

Fig. 8-10 Large shallow plate load test

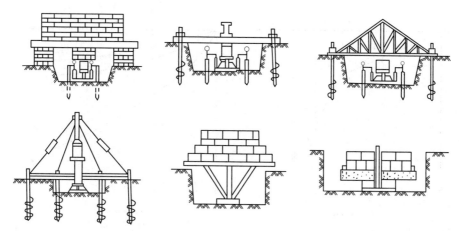

Fig. 8-11 Several common load test equipment

and leveling instrument. The common load test equipment is shown in Fig. 8-11.

② Basic requirements of load test. Load tests should be arranged in a representative place, and the number of tests is at least 3, and more when the site is uneven. The shallow load test should be arranged at the elevation of the foundation bottom.

The grading loading method of settlement stability shall be (conventional slow method). If there is comparative experience, in order to speed up the test, the grading loading method of settlement instability (fast method) or equal settlement rate method can also be used. The number of loading grades should be 10~12 and no less than 8. The measurement accuracy of load value should not be less than ±1% of the maximum load. The settlement of pressure plate can be measured by percentage meter or electric displacement measurement, and its accuracy should not be less than ± 0.01 mm.

For the slow speed method, when the test object is soil, the settlement is measured once at intervals of 5 min, 5 min, 10 min, 10 min, 15 min and 15 min after each load is applied, and then the settlement is measured once at intervals of 30 min. When the settlement is less than 0.1 mm per hour for two consecutive hours, it can be considered that the settlement has reached the relative stability standard and the next load shall be applied. When the test object is rock mass, after each stage load is applied, the settlement is measured at intervals of 1 min, 2 min, 2 min and 5 min, and then the settlement is measured at intervals of 10 min. When the reading difference is less than 0.01 mm for three consecutive times, it can be considered that the settlement has reached the relative stability standard and the next stage load shall be applied.

The test should be carried out until up to the failure stage. When one of the following cases occurs, it can be considered that the foundation soil has reached the limit state, and the load test can be terminated: (a) There is obvious lateral extrusion in oil around the bearing plate, and obvious uplift or continuous radial cracks in the surrounding rock and soil. (b) The settlement of this stage load is 5 times larger than that of the previous stage load, and there is obvious steep drop in load and settlement curve. (c) The settlement rate of 24 h under a certain load cannot reach the relative stability standard. (d) The ratio of the total settlement to the diameter (or width) of the bearing plate exceeds 0.06.

③ Application of load test results

(a) Determine the bearing capacity of foundation to provide the basis for evaluating the bearing capacity of foundation soil.

According to the p (load)-s (settlement) curve and s (settlement)-t (time) curve

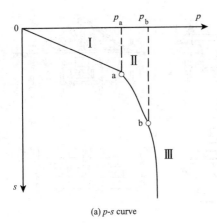
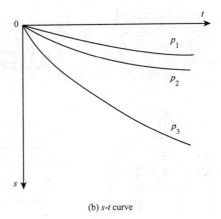

(a) *p-s* curve (b) *s-t* curve

Fig. 8-12 Typical load test curves

obtained from the load test (Fig. 8-12), the bearing capacity of the foundation can be determined according to the methods in Appendix C and D of *Code for Design of Building Foundation* (GB 50007—2011).

(b) Determine the deformation modulus E_0 of foundation soil and the reaction coefficient of foundation soil, which can be calculated according to the relevant values on the p-s curve with relevant formula.

2) Static cone penetration test (CPT)

① Equipment of static cone penetration test. Static penetrometer is generally composed of three parts. Penetration system includes a pressure device and a reaction device, whose function is to press the probe into the soil layer at a constant speed and vertically. The measuring system is used to measure and record the resistance of the probe. There is a resistance sensor in the static contact probe, and it transmits the penetration resistance through the signal and mechanical system to the automatic recorder and draws the resistance curve with depth (Fig. 8-13).

② Application of CPT results. According to the measurement results of static cone penetration test, the following results can be obtained: specific penetration resistance-depth $(p_s\text{-}h)$ curve, cone tip resistance-depth $(q_c\text{-}h)$ curve, side wall friction resistance-depth $(f_s\text{-}h)$ curve and friction ratio-depth $(R_f\text{-}h)$

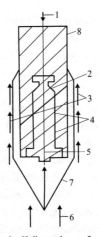

1—Empowerment; 2—Hollow columns; 3—Side wall friction; 4—Resistors; 5—Pillars; 6—Conical tip resistance; 7—Probe sleeve; 8—Probe tube

Fig. 8-13 schematic diagram of working principle of touch probe

curve. For the pressure probe, we can also get the pressure-depth $(u\text{-}h)$ curve, as shown in Fig. 8-14.

The application of CPT results mainly has the following aspects:

(a) Classifying soil horizon. According to the curves obtained by the static cone penetration test, the soil layer can be divided according to the similar q_c and R_f. For the pore pressure probe, the pore water pressure can also be used to divide the soil layer.

(b) Estimation of physical and mechanical properties of soil. According to a large number

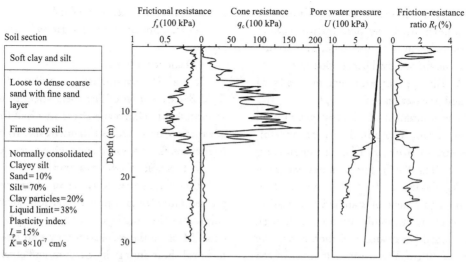

Fig. 8-14 The results of static cone penetration test curve and corresponding soil profile

of experimental data analysis, the relationship between undrained shear strength c_u and q_c of cohesive soil, the relationship between specific penetration resistance p_s and compression modulus E_s and deformation modulus E_0 of soil, the consolidation coefficient of saturated clay and the density of sand can be obtained. Please refer to relevant manuals and specifications as many departments at home and abroad have put forward many practical formulas.

(c) Determining the bearing capacity of shallow foundation. According to the specific penetration resistance p_s of static cone penetration test, the bearing capacity of shallow foundation can be determined by empirical formula.

(d) Estimating bearing capacity of single pile. Using the results of static cone penetration test to estimate the bearing capacity of pile, and there are some mature empirical formulas applied in China.

(e) Determination of liquefaction potential of saturated sand and silt. Saturated sand and silt may be liquefied under earthquake, against which static cone penetration test can be used to judge liquefaction. Such test has the advantages of continuous, fast, high efficiency, multi-function, and the dual role of exploration and test with high accuracy and good reproducibility. The static cone penetration test is suitable for cohesive soil and silty soil, and loose to medium dense sand, but its disadvantage is that it is difficult to penetrate gravel soil and dense sand, and soil layers cannot be observed directly.

3) Dynamic penetration test

There are mainly two categories of dynamic penetration test, namely cone dynamic penetration and standard penetration test (SPT). The common point is to use a certain hammer kinetic energy to penetrate a specification of the probe into the soil, and determine the property of the soil according to the energy required for a certain depth of each penetration, and layered the soil. The required energy reflects the resistance of soil, which can be generally expressed by the number of hammers.

According to the hammer energy, cone penetration can be divided into light (hammer weight 10 kg), heavy (hammer weight 63.5 kg) and super heavy (hammer weight 120 kg). The difference between standard penetration test and dynamic penetration test is mainly that cone penetration contact probe is not conical, but cylindrical probe with standard specifications, which is composed of two semicircular tubes, commonly known as the penetrator. Cone penetration adopts intermittent penetration method. This section focuses on the standard penetration test.

① Equipment and test methods of standard penetration test

The equipment of standard penetration test is mainly composed of a penetrator, a penetrator probe rod and a penetrating hammer (Fig. 8-15). The hammer weight is 63.5 kg. Under the free fall distance of 76 cm, the equipment of standard penetration test penetrates the soil layer 15 cm through a cylindrical penetrator, and then enters the depth of 30 cm. The hammer number of the later 30 cm is called the number of standard penetration test, which is expressed as $N_{63.5}$. The influencing factors include drill pipe length, drill pipe connection mode and so on, so sometimes it's necessary to correct length of $N_{63.5}$.

② Application of standard penetration test results

(a) Classification of soil or soil profile;

(b) Determination of sand compactness and seismic liquefaction;

(c) To judge the consistency state of cohesive soil and the values of c and φ;

(d) Evaluation of deformation modulus E_0 and compression modulus E_s of soil;

(e) Determination of foundation bearing capacity.

4) Vane shear test

It is a simple and reliable in-situ test for rapid determination of fast shear strength of saturated soft clay. Vane shear test has the advantages of small soil disturbance, light equipment, fast test speed and high efficiency.

① Principle of vane shear test

Torque is applied to the cross-plate head pressed into clay to form a cylindrical failure surface in the soil layer in the cross-plate head. Measure the maximum torque to resist torsional shear during shear failure, and obtain shear strength of soil by calculation (Fig. 8-16).

② Scope of application and results application

The vane shear test can be used to determine the undrained shear strength and sensitivity of saturated soft clay ($\varphi \approx 0$).

Application of test results: determine

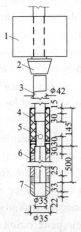

1—Cardiac hammer; 2—Hammer pads; 3—Probe rods; 4—Insertion head; 5—The outlet hole; 6—Insert body; 7—Penetration boots

Fig. 8-15 Standard penetration test facility (mm)

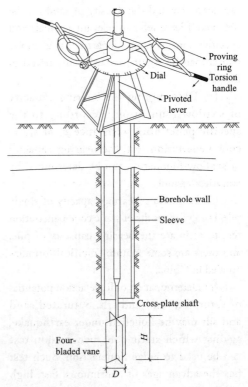

Fig. 8-16 Vane shear test

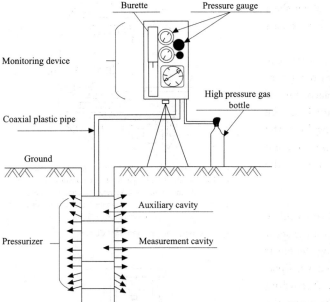

Fig. 8-17 Pressuremeter test

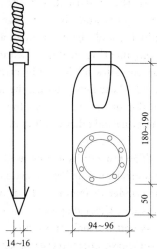

Fig. 8-18 Shovel side dilatometer

the foundation bearing capacity, bearing capacity of single pile, calculate slope stability, determine the consolidation history of soft clay soil.

5) Pressuremeter test (PMT)

Fig. 8-17 is the schematic of pressuremeter test. Compared with the static load test, the pressuremeter test has the characteristics of high precision, light equipment and short test time, but its accuracy is greatly affected by the hole quality.

6) Flat dilatometer test (DMT)

Flat dilatometer test (referred to as flat dilatometer test for short) is to penetrate a flat spatula probe (Fig. 8-18) into the soil by static force (sometimes hammering power), and after the probe reaches the test depth, use air pressure to make the round steel membrane on the side of the flat spatula expand outward for testing. It can be used as a special side pressure test. The advantages of flat dilatometer test are simplicity, rapidity, good repeatability and low cost.

(2) Field tests on mechanical properties of rock mass

The test purpose, technical requirements, test methods and steps, result summarizing and application of the static load test of rock mass in plate load test have been explained in the previous section "Static load test of soil mass", so they are not repeated here.

1) Rock strength test

The shear test instrument of rock mass is composed of three systems: loading, force transmission and measurement (Fig. 8-19). The principle of field direct shear test is basically the same as that of indoor direct shear test. However, because the rock mass of the test method is much larger than the indoor sample, it can include the change of macro structure, and the test conditions are close to the in-situ conditions, the results are closer to the actual engineering situation.

① Types of in-situ direct shear tests of rock mass

In-situ direct shear test of rock mass can be divided into three types: shear test of rock mass itself, shear test of rock mass along weak structural plane and shear test of rock mass and concrete interface. Further, it can be divided into shear test of rock mass specimen failure along shear plane under normal stress, shear test of rock mass after shear along shear plane (friction test) and shear test of rock mass when normal stress is zero.

In the in-situ direct shear test, the test objects and the corresponding test methods should be determined according to the engineering geological conditions, the characteristics of the engineering load, the possible shear failure mode, the position and direction of the shear plane, the stress of the shear plane and other conditions.

② Test results

The test results can be used to calculate the normal stress and shear stress on the shear plane under different loads, draw the shear stress and shear displacement curve, shear stress and normal stress curve. According to the curve characteristics, determine the proportional strength, yield strength, peak strength, dilatancy point and dilatancy strength of rock mass, and the corresponding values of c and by Coulomb expression.

2) Point load strength test

The main equipment is the point load meter, which is composed of pressure system (including oil pump, pressure bearing, frame, jack and conical spherical pressure head) and pressure gauge (Fig. 8-20). There are five sample loading modes: radial, axial, irregular, vertical or parallel structural planes. Place the rock sample between the two loading cones of the point load meter, slowly and uniformly pressurized until the rock sample ruptures. Record the failure load and measure the cross-section area of failure plane.

8.2.4 Overview of Laboratory Tests

There are many methods of laboratory testing, which can be divided into the following categories:

(1) Test of physical properties of soil

Sandy soil: particle size distribution, relative density (specific gravity), natural water content, natural density, maximum and minimum dry density.

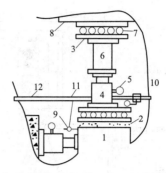

1—Rock mass specimen; 2—Cement mortar; 3—Steel plate; 4—Jack; 5—Pressure gauge; 6—Force transmission column; 7—Roller group; 8—Concrete; 9—Dial indicator; 10—Surrounding rock; 11—Magnetic watch connector frame; 12—U-shaped steel beam

Fig. 8-19 Rock mass shear test device

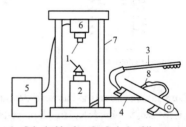

1—Spherical loaders; 2—Jack; 3—Oil pumps; 4—High pressure hose; 5—Four pressure gauge indicator; 6—Pressure sensor; 7—Frame; 8—Quick high pressure

Fig. 8-20 Digital display point load meter

Silt: particle gradation, liquid limit, plastic limit, relative density, natural water content, natural density and organic matter content.

Cohesive soil: liquid limit, plastic limit, relative density, natural water content, natural density and organic matter content.

(2) Soil compression and consolidation test

(3) Shear strength test of soil

Such as direct shear test, various conventional triaxial specimens, unconfined compressive strength test.

(4) Experiment on dynamic properties of soil

Such as dynamic triaxial test, resonance column test, dynamic simple shear test.

(5) Rock test

Such as rock and mineral identification, block density test, water absorption rate and saturated water absorption rate test, disintegration resistance test, expansion test.

8.2.5 In-situ Inspection and Monitoring

(1) In-situ inspection

The inspection work should include two aspects: First, check and verify the engineering geological investigation results and evaluation suggestions, that is, the first-hand engineering geological and hydrogeological data obtained by exposing rock and soil through foundation pit excavation during construction. The data are more accurate than those obtained in the investigation stage, and can be used to supplement and revise the investigation results. If there is a big difference between the actual situation and the investigation results, make supplementary investigation or modification of geotechnical engineering design, and take corresponding treatment measures in the construction stage; Second, control and inspect the geotechnical engineering construction quality, namely construction supervision and quality control, such as the inspection of foundation trench in building engineering, a series of quality monitoring in pile foundation construction, monitoring of foundation treatment construction quality, monitoring of construction quality of deep foundation pit support system.

(2) Long-term monitoring

Long-term monitoring refers to the observation of adverse geological phenomena, rock and soil properties and groundwater affecting the project during the construction process and operation. The purpose is understanding the degree of influence caused by the construction, monitoring geological changes and development rules, so as to take appropriate prevention and control measures in the design and construction in time, to ensure the safety of the project. Monitoring work mainly includes three aspects. First, the monitoring of rock and soil properties under load, for example, deformation and displacement monitoring of rock and soil, stress measurement of rock and soil. Second, the monitoring of building structures in construction or operation. For particularly important buildings such as nuclear power plants, monitoring should be carried out throughout the operation period. Third, the monitoring of environmental conditions. It includes the monitoring of certain factors in engineering geological conditions, especially adverse geological phenomena (such as landslides, collapses, mudslide, soil caves, etc.) that can pose a threat to projects. Measurement points should be arranged to carry out corresponding long-term monitoring work from the beginning of the investigation period done.

8.3 Report of Engineering Geological Investigation

The investigation report generally consists of two parts: text and diagrams.

(1) The basic content of the investigation report

Different project types have different investigation requirements, and the focus of the report is also different.

Report generally consists of several parts: introduction, general description, monograph and conclusion.

The introduction part mainly explains the tasks of the investigation work, the working methods adopted and the results obtained, etc.

The general description mainly clarifies the engineering geological conditions of the construction area, and the regional natural geography and geological background.

The monograph is the central content of

the entire report. It mainly demonstrates the engineering geological issues, i.e., engineering geological conditions and engineering geological problems involved in the construction, evaluates the suitability of the project, and proposes measures to prevent and control adverse geological effects.

The conclusion is to give brief concluding opinions on various specific issues on the basis of monographs.

(2) Charts that should be attached to the report

The investigation report should be attached with necessary charts, mainly including:

1) Engineering geological plan (house building with investigation engineering layout);

2) Borehole histogram or comprehensive engineering geological histogram;

3) Longitudinal and transverse sections of engineering geology;

4) Chart of laboratory test and in-situ test results.

Quiz

1. What are the specific basic tasks of engineering geological investigation?

2. Briefly describe the types of methods and technical means used in engineering geological investigations.

3. How are the stages of engineering geological investigation divided?

4. What is engineering geological investigating and mapping? Briefly describe its content and methods.

5. What are the main methods for engineering geological prospecting?

6. What is the purpose and task of drilling? What procedures are involved in the drilling process?

7. How many grades of soil sample quality can be divided? What are the requirements for sampling tools and sampling methods for each level?

8. What is geotechnical in-situ testing? What are its advantages and disadvantages?

9. What is the basic principle of static load test? What are its technical requirements?

10. What is the purpose and principle of static sounding? What are the main uses of the results?

11. What is the purpose and principle of the standard penetration test? What are the application aspects of the results?

12. What is the purpose and applicable conditions of the cross-plate shear test? Which physical and mechanical property parameters of the soil can be obtained?

13. What is the purpose of the side pressure test? What are the main results?

14. What are the methods of rock mass in-situ test, and what is its main principle?

15. Briefly describe the basic contents and attached drawings of geotechnical engineering investigation report.

References

[1] BIENIAWSKI Z T. Engineering rock mass classifications [M]. New York: The Wiley-Interscience Publication, 1989.

[2] CAI M F, HE M C, LIU D Y. Rock mechanics and engineering [M]. Beijing: Science Press, 2002.

[3] CHE Y T. On the current situation of earthquake prediction and its basic research problems[J]. International earthquake dynamics, 2005, 12: 19-23.

[4] CHEN B R, HONG Z J, WANG F X. Groundwater dynamics and prediction [M]. Beijing: Science Press, 1988.

[5] CHEN L X, CHEN S G, TU P, et al. Study on mutual relationships between surrounding rock classifications by Q value, RMR and BQ method for underground cavern[J]. Subgrade Engineering, 2017, (06): 107-112.

[6] Chinese geological disaster prevention engineering industry association. Standard of classification for geological hazard, T/CAGHP 001—2018 [S]. Beijing: China University of Geosciences Press, 2018.

[7] Editorial Committee of Engineering Geology Handbook. Engineering geology handbook (Fifth edition) [M]. Beijing: China Architecture & Building Press, 2018.

[8] BELL F G. Engineering geology (Second Edition) [M]. Burlington: Elsevier Press, 2007.

[9] FAN X R. Simplified graphical method of basic quality index BQ in "Standard for Engineering Classification of Rock Mass" [J]. Water Resources and Hydropower Technology, 2016, 47(5): 154-156, 170.

[10] GAO D Z. Geotechnical engineering survey and design [M]. Beijing: China Communications Press, 2010.

[11] GAO J C, DU G Y. Geotechnical engineering investigation and evaluation[M]. Wuhan: China University of Geosciences Press, 2003.

[12] GAO W, LIU Q S. Back analysis of underground engineering based on bionic computational intelligence-theory and application[M]. Beijing: Science Press, 2009.

[13] HU H T. Collapse and rockfall [M]. Beijing: China Railway Publishing House, 1989.

[14] HUANG R Q. A typical catastrophic landslide in China[M]. Beijing: Science Press, 2008.

[15] HUANG R Q. The mechanism of landslide disaster triggered by the Wenchuan M8.0 earthquake and its geo-mechanical model[J]. Chinese Journal of Rock Mechanics and Engineering, 2009, 28(6): 1239-1249.

[16] HUANG R Q, LI W L. Research on the development and distribution of geological hazards triggered by the "5.12" Wenchuan earthquake[J]. Chinese Journal of Rock Mechanics and Engineering, 2008, 27(12): 2585-2592.

[17] JIANG J H, XU G B. Engineering geology [M]. Chengdu: Chengdu University of Science and Technology Press, 1995.

[18] JIANG J S, XU Y, LIANG H Q. Engineering geology [M]. Beijing: China Architecture & Building Press, 2016.

[19] KONG S L. Engineering Geology (Second edition) [M]. Chongqing: Chongqing University Press, 2005.

[20] KONG X L. Engineering geology

[M]. Beijing: China Architecture & Building Press, 1997.

[21] LI X R. Engineering geology[M]. Beijing: China Electric Power Press, 2016.

[22] LI Z J, JIN A W, WEI J C. Engineering geology [M]. Beijing: Chemical Industry Press, 2018.

[23] LIU X R, YANG Z P. Engineering geology[M]. Beijing: China Machine Press, 2021.

[24] LI Z Y, YANG Y Y. Introduction to engineering geology [M]. Wuhan: China University of Geosciences Press, 1994.

[25] LIU Z C, LI G H, ZHU K. Water supply hydrogeology [M]. Beijing: China Architecture & Building Press, 1998.

[26] LIU Z Y. Hydrogeological manual [M]. Changchun: Yinsheng Audio Visual Publishing House, 2010.

[27] LIU Z Y, ZHU Y Z, XIAO Z R, et al. Engineering geology (Second edition) [M]. Beijing: China Electric Power Press, 2016.

[28] FREITAS M H. Engineering geology-principles and practice[M]. New York: Springer Press, 2009.

[29] Ministry of Construction of the People's Republic of China. Code for investigation of geotechnical engineering, GB. 50021—2001 [S]. Beijing: China Architecture & Building Press, 2009.

[30] Ministry of Housing and Urban-Rural Development of the People's Republic of China. Standard for engineering classification of rock mass, GB/T 50218—2014 [S]. Beijing: China Planning Press, 2014.

[31] Ministry of Water Resources of the People's Republic of China. Standard for geotechnical testing method, GB/T 50123—2019 [S]. Beijing: China Planning Press, 2019.

[32] Ministry of Housing and Urban-Rural Development of the People's Republic of China. Standard for building construction in collapsible loess regions: GB 50025—2018 [S]. Beijing: China Planning Press, 2019.

[33] NIU Y N. Engineering geology [M]. Beijing: Chemical Industry Press, 2016.

[34] PRICE D G. Engineering geology-principles and practice[M]. Berlin: Springer, 2009.

[35] SHANG M. Engineering geology [M]. Beijing: Geological Publishing House, 2017.

[36] SHEN Z L, LIU G Y, YANG C T, et al. Hydrogeology [M]. Beijing: Science Press, 1985.

[37] SHI B, YAN C H. Engineering geology [M]. Beijing: Science Press, 2019.

[38] SHI Z M, KONG X L. Engineering geology (Second Edition) [M]. Beijing: China Architecture & Building Press, 2011.

[39] SUN J Q. Engineering geology (Third edition) [M]. Wuhan: Wuhan University of Technology Press, 2007.

[40] TAO X F, WU D C. General geology[M]. Beijing: Science Press, 2007.

[41] TONY W. Foundations of engineering geology (Third edition) [M]. London: Spon Press, 2009.

[42] WANG D, ZHANG R M, SHI Y L, et al. Basic of hydrogeology[M]. Beijing: Geological Publishing House, 1995.

[43] WANG G L. Engineering geology [M]. Beijing: China Architecture & Building Press, 2012.

[44] WANG K H. Geotechnical engineering investigation (Second edition) [M], Beijing: China Architecture & Building Press, 2016.

[45] WU A Q, WANG B. Engineering rock mass classification method based on rock mass quality index BQ for rock slope [J]. Chinese Journal of Rock Mechanics and Engineering, 2014, 33(4): 699-706.

[46] WU F Q, QI S W. Engineering geology: science, art, and challenges—The development of engineering geology from the 2014 national annual meeting of engineering geology[J]. Journal of Engineering Geology, 2015, 23(1):1-6.

[47] XIANG W, TANG H M. Geotechnical engineering investigation[M]. Beijing: Chemical Industry Press, 2012.

[48] XU Q, PEI X J, HUANG R Q. Research on large-scale landslides in Wenchuan earthquake[M]. Beijing: Science Press, 2009.

[49] YANG Z P, HU Y X, HUANG D. Engineering mechanical properties of earth-rock mixture and stability of ultra-high fill[M]. Chengdu: Sichuan University Press, 2018.

[50] ZHANG R Q, LIANG X, JIN M G, et al. Basics of hydrogeology[M]. Beijing: Geological Publishing House, 2010.

[51] ZHANG R Q, LIANG X, JIN M G, et al. Foundation of hydrogeological (Seventh Edition) [M]. Beijing: Geological Publishing House, 2018.

[52] ZHANG Z M. Engineering geology[M]. Beijing: China Architecture & Building Press, 2007.

[53] ZHANG Z Y, WANG S T, WANG L S, et al. Principles of engineering geological analysis (Fourth Edition) [M]. Beijing: Geological Publishing, 2009.

[54] ZHOU R J, HUANG R Q, LEI J C, et al. Surface rupture and seismic damage characteristics of the WenchuanMS8.0 earthquake in Sichuan[J]. Chinese Journal of Rock Mechanics and Engineering, 2008, 27(11): 2173-2183.

[55] ZHOU W Y, YANG Q. Numerical calculation method of rock mechanics[M]. Beijing: China Electric Power Press, 2005.

[56] ZOU C, Yan L M. New geological English[M]. Beijing: Publishing House of Electronics Industry, 2013.